MW00444783

VISION

A Holistic Guide to Healing the Eyesight

by Joanna Rotté, Ph.D.
and Koji Yamamoto

Japan Publications, Inc.

Illustrations by Anna Demchick

Note to the reader: The information contained in this book is not intended to be used in the diagnosis, prescription, or treatment of disease or any health disorder whatsoever. Nor is this information intended to replace competent medical care. This book is a compendium of information which may be used as an adjunct to a rational and responsible health care plan.

Published by JAPAN PUBLICATIONS, INC., Tokyo and New York

Distributors:
UNITED STATES: *Kodansha International/USA, Ltd., 114 Fifth Avenue, New York, N. Y. 10011.* CANADA: *Fitzhenry & Whiteside Ltd., 195 Allstate Parkway, Markham, Ontario, L3R 4T8.* MEXICO AND CENTRAL AMERICA: *HARLA S. A. de C. V., Apartado 30–546, Mexico 4, D. F.* BRITISH ISLES: *Premier Book Marketing Ltd., 1 Gower Street, London WC1E 6HA.* EUROPEAN CONTINENT: *European Book Service PBD, Strijkviertel 63, 3454 PK De Meern, The Netherlands.* AUSTRALIA AND NEW ZEALAND: *Bookwise International, 54 Crittenden Road, Findon, South Australia 5007.* THE FAR EAST AND JAPAN: *Japan Publications Trading Co., Ltd., 1–2–1, Sarugaku-cho, Chiyoda-ku, Tokyo 101.*

First edition: April 1986
Second printing: August 1989

LCCC No. 84–80538
ISBN 0–87040–622–1

Printed in U.S.A.

Foreword

The application of a balanced state of mind and vision as taught in this text should not be surprising or novel to the reader. However, due to the promotion of eyeglasses and/or contact lenses as a treatment for myopia (nearsightedness) many of the concepts in this book may appear new to the audience. Although he may not have been the first to describe the relationship of a state of mind and vision, the very well-known and highly regarded philosopher and physician Moses ben Maimon (Maimonides) wrote in the twelfth century:

> You must consider, when reading this treatise, that mental perception, because connected with matter, is subject to conditions similar to those to which physical perception is subject. That is to say, if your eye looks around, you can perceive all that is within the range of your vision; if, however, you overstrain your eye, exerting it too much by attempting to see an object which is too distant for your eye, or to examine writings or engravings too small for your sight, and forcing it to obtain a correct perception of them, you will not only weaken your sight with regard to that special object, but also for those things which you otherwise are able to perceive: your eye will have become too weak to perceive what you were able to see before you exerted yourself and exceeded the limits of your vision (Moses Maimonides, *The Guide for the Perplexed* [New York: Dover Publications, Inc., 1956]).

These words by Maimonides are echoed by the authors of this book on page 137:

> If you can train yourself to see without struggling to see, you will avoid eyestrain and your power of vision will become stable.

A moment should be taken on the consideration of this very important message concerning the state of mind and vision. Whether adhering to Eastern or Western philosophies, there should be a firm understanding of the role of the nervous system, and in particular the autonomic nervous system, in regard to the state of well-being of an individual. The practical application of this philosophy is well described in this book as being a balance between the mind and the body. The authors stress the dynamic relationship between diet, breathing, exercise, and the state of mind as the important factors in vision. The reason that these factors are important is that they contribute to the condition of the autonomic nervous system —that part of the nervous system that controls bodily functions. The autonomic nervous system is comprised of two components: the parasympathetic and the sympathetic. The two components act as regulators for our bodily functions. For example, the sympathetic nervous system constricts the peripheral blood vessels, while the parasympathetic nervous system dilates peripheral blood vessels. The sympathetic nervous system causes dilation of the pupil of the eye, while the parasympathetic nervous system causes constriction of the pupil of the eye. The sympathetic nervous system causes dilation of the ciliary muscle of the eye, while the parasympathetic nervous system causes constriction of the ciliary muscle of the eye. The ciliary muscle is commonly known as the focusing muscle of the eye, since through its action we are able to focus on objects at different distances in space. The ciliary muscle is a ring muscle, that can either dilate or constrict. Attached to the muscle are small thread-like fibers, the zonules of Zinn, which in turn are attached to the lens inside our eyes. As the ciliary muscle constricts (parasympathetic activity), the tension on the zonules is relaxed, thereby allowing the lens to become more curved. This increases the focusing power of the eye to allow us to see objects that are near. As the ciliary muscle dilates (sympathetic activity), the tension on the zonules is increased, thereby allowing the lens to become less curved. This decreases the focusing power of the eye to allow us to see objects that are far. The process of changing the focus of the eye is known as accommodation.

The delicate balance between the sympathetic and the parasympathetic innervation to the ciliary muscle determines the status of the lens of the eye. With chronic overactivation of the parasympathetic nervous system, we note a chronic increase in the curvature of the lens of the eye—clinically this is known as myopia. Recent research evidence, beginning with Neal Miller's article: "Learning of Visceral and Glandular Responses" (*Science*, January 1969), shows us that we can develop inappropriate autonomic nervous system learning due to such factors as unpleasant emotional experience—what we usually call stress. The development of myopia is typically an example of such inappropriate autonomic nervous system learning.

The reading of this text should allow the reader to have an increased understanding of the relationship between the state of mind and vision. While the techniques and training methods described by the authors may seem lengthy in practice and one's progress may sometimes appear too slow, that is the nature of breaking old, bad habits. However, once the proper balance is achieved, the reader will find the new way of vision rewarding and will be careful not to revert to the old, bad habits. The balance may seem elusive at times—since if we try too hard we may not be successful. I would like to conclude with these summary comments from the well-known book, *The Art of Seeing* by Aldous Huxley (California: Creative Arts Book Company, 1982):

> When we see, our minds become acquainted with events in the outside world through the instrumentality of the eyes and the nervous system. In the process of seeing, the mind, eyes and nervous system are intimately associated to form a single whole. Anything which affects one element in this whole exer-

cises an influence upon the other elements. In practice, we find that it is possible to act directly upon the eyes and the mind. The nervous system which connects them cannot be influenced except indirectly.

September 4, 1985

Joseph N. Trachtman, O.D., Ph.D.
57 Hicks Street
Brooklyn Heights, New York 11201

Preface

When looking around on the streets of cities or out of a car window onto suburban avenues or into the faces of people shopping, I see human beings in stages of anguish, bodies uncomfortable, and insecurity or confusion flickering through lackluster eyes. Adults in general have the appearance of being in trouble, which is not said to speculate in media-like morbidity nor to protract the literary dissatisfaction with contemporary times, but is said in hopes of aiming toward a future when troubles might be lessened, when much of human suffering is recognized as unnecessary, at least the repetition of it, and avoidable.

In the rather middle-income building housing four residential units where I was renting an apartment, there lived five people in addition to my son and I. All wore glasses and endured problems. The eldest in this building, a woman of eighty-five, had undergone a cataract operation on each eye, wore very thick glasses, feared the power of sunlight, dreaded going outdoors, and was an exceedingly lonely person who frequently cried, consistently prayed, and too often in a day rang at my door with an offering of food in exchange for a greeting. Regularly she took in her several cups of weak coffee teaspoons of white sugar, and she ate white bread, white pasta, simulated cheeses, canned vegetables and packaged pastries. People who visited our building would exclaim how fine it is that a woman so elderly could yet perform housework on her knees, but I disagreed. It seemed to me that as long as was the quantity of her years was the depth of her sadness. Her son, an on-again/off-again odd-job, overweight bachelor in his fifties, for whom she cleaned, laundered and labored except on Sundays, had a habit of blinking excessively and an inability to countenance another's eyes when conversing. In the next-door apartment on the upper level resided a widow of seventy-nine, myopic and presbyopic, and so hump-backed that when out walking her immediate range of vision encompassed only the ground. The rooms beneath her were occupied by a married couple, working professionals in their late thirties, both wearers of eyeglasses, and both survivors of spinal surgery. Those were my neighbors. Possibly none of their conditions could ever exemplify the standard of average American health, but would rather be judged as American sub-health. Yet there they were, my five surrounding neighbors, and all of them were ailing people with failing or failed eyesight.

More optimistically, the physical condition of the average American might be characterized by that of any of the members of the aerobics class in which I participated three mornings a week for the past year. Mostly women, these jogging, dancing, moving, attractive beings were decidedly more fit than my neighbors. Still, of the usual twelve or so in attendance, two exercised with glasses on, half a dozen ran poorly, and all but three were slow to start and not hearty at the finish. Interestingly, the women chatted while jogging, implying a desire to avoid experiencing the body in motion and the mind in tune. Socialization was more the issue of the class than deep breathing, and unconscious repetition of movement allowed to become routine was its vehicle. Real running, the kind of running that lets the runner feel the blood circulating and the nervous system being brought to a peak, rarely revealed itself. Only three people, less than a quarter of the membership, ever seemed to exercise with an eye toward stretching limitations, toward freedom, and one of those was the teacher, who ran wearing contact lenses.

A few years ago I had the charming good fortune to interview Moshe Feldenkrais, the movement specialist. A no-nonsense, grandfatherly personality and a meticulous thinker, Mr. Feldenkrais mused how throughout America a great number of men and women are jogging—but to what effect, he questioned, because with what kind of posture is everyone running, with what manner of imbalance, for example, in the shoulder blades and pelvis. He said, "There are 25,000,000 people jogging in America. Why do they jog? Because they feel bad. If they were feeling well, they wouldn't be jogging. And by the way, their jogging is not much good either. There are very few people who jog and improve." In his comment on lack of improvement, Mr. Feldenkrais was surely not talking about failing to meet target heart rates (which joggers do meet and even surpass), but about a more profound lack of improvement, the absence of a felt sense of well-being. The validity of his criticism can be strikingly borne out by observing the runners in the Boston (or any other) Marathon. An astounding percentage of the entrants exert themselves admirably to fulfill the twenty-six miles, but with such tension in the upper body, especially in the neck and around the shoulders and the jaw, that it seems to scream out for release.

Would we not consider it extraordinarily abnormal to come upon a zebra in the veldt or a llama in the mountains with a stiff neck, or a panther with myopia, or a lion with ulcers? How weird, unsightly, or just pitiful would we consider an elephant with a twist in the hips or a squirrel with tension between the eyes? Yet we accept as normal the tight, distorted or anxious postures so common among human beings seated around tables, in movie theaters, behind desks, or standing in factories, or even working-out in gymnasiums. Furthermore, we look outside ourselves—to heredity, viruses, or the effects of society and the environment upon us—to explain our troubles. How frequently do we look inside ourselves—to our use of the body we inherited, to the strength of our life-force for resisting or assimilating alien agents, to our responses to the stresses of society and the artificiality of the environment, in short, to the components of our life-style—to solve the source of our discomforts? Perhaps our way of looking at ourselves, and not only our physiological vision, is frightfully shortsighted.

One intention of this book is to render unacceptable handicaps and abnormalities that need not persist as handicaps and abnormalities. In spite of the passive attitude toward the epidemic amount of nearsightedness in America today and in spite of the massively widespread use of corrective lenses as a supposed cure for nearsightedness, this book was written, fundamentally, so that any sustained impairment of a human being's eyesight might be recognized for what it is: a chronic disease; and so that the real causes of an individual's eye disease might be pinpointed by him or her and the natural means to recover his or her essential good vision might be determined and undertaken without suffering.

How unseemly, how much less than fully human, to enter one's grave without having seen the truth of one's condition and without having ever taken one's condition in hand and changed it, radically adjusted it, of one's own volition and by one's own power, changing, for example, one's eyesight from poor to sharp, or, greater yet, one's way of seeing from deluded to illuminated! Both can of course be accomplished. The shape of the eyeball can be changed, as can the shape and entire functioning of the whole body. So can the perspective of the mind be changed.

This book is, above all, about change. Based on excerpts from the life story and the teachings of Masahiro Oki (with whom Koji Yamamoto and I studied in Japan—he for more than fifteen years), it is given to help people help themselves. If I were to consider the contents of this book in terms of the problems of my recent neighbors, I would say that an application of the book's principles and various techniques could bring relief, even rejuvenation, to them all. But, to qualify, I must admit that my recent neighbors, perhaps with the exception of the married couple, would be disinterested in, or afraid of, the self-help, even self-liberation thrust of the book. The book is best suited, I would suggest, to the adventurous.

Koji Yamamoto, for his part, wishes the reader to use the book as a guide, a tool, for practice by oneself or with friends or family, rather than as a text simply to be read; and to take from it those techniques that are appropriate and applicable to oneself. I second his counsel, and, in addition, want the reader to know that my own eyesight is less than perfect (to cure it was the primary reason I went to Japan). But the reader should also know that my eyesight has been worse, as it has been better, than it is now, and I look forward to its becoming better yet again. In other words, I look forward to continuing to change myself, and wish the same for the reader, through application of the principles and selected techniques of this book. Change is, after all, a process on-going, life-long.

The ideology and methodology of this book are most indebted to the wisdom of Masahiro Oki, late master teacher of Zen and Yoga in Japan, while any faults in its organization or presentation are entirely the responsibility of the authors, due to their own limitations, and are not at all to be attributed to Mr. Oki. For gracious contributions to the completion of this book, the authors would like to thank Karen Kraruse for her fine photography and Annie Demchick for her excellent illustrations; Dana Geller, David Pavlosky and Gini Schwartz for their energy and exquisite form in a difficult modeling assignment; Eli Corin, Margot Jones and Susan Kaner for their work in the preparation of the manuscript; and Yoshiro Fujiwara and Iwao Yoshizaki of Japan Publications, Inc. for their editorial assistance. With gratitude and deep respect, the authors dedicate this book to Masahiro Oki sensei.

Joanna Rotté
Rosemont, PA

Contents

Chapter 3: Health and Eye Disease 55

Chapter 4: Holistic Vision 67

Introduction

Nearsightedness Can be Improved

It is commonly held throughout the modern world, and it is assumed by Western medicine, that myopia, or nearsightedness, cannot be cured. But from the experiences related in this book you may realize that myopia cannot only be improved but recovered completely.

At this time you might think of nearsightedness as strictly a physical disorder. Yet, in deference to traditional wisdom, you could surely concede that the eyes are the windows of the mind. The process of seeing involves the mind much more than we may possibly imagine, and, in truth, the primary cause of nearsightedness is perpetual mental and physical tension.

First, it needs to be understood that deficient eyesight is a chronic disease, indicating a lifelong malfunctioning of the body and mind. No vision problem is limited to some defect in the eyes only. Please grasp the concept that all parts of the body work interdependently and that the body and mind are integrated.

To improve any illness it is vitally important to look into oneself and not outside oneself for a recovery; that is, to rid oneself of a dependent mentality. The body has its own preservative and healing powers, originating from the life-force. Unfortunately, in daily living you might be oppressing the healing power of the life-force by being heedless of your posture and eating habits and by being unaware of the nature of your own mentality and spirituality. All disease, unbalanced vision included, can actually be interpreted as a warning, a message from the life-force saying that the life-style is unbalanced.

Yoga for Self-Healing

Yoga is a way of balancing the life-style. Even though it originated in India four to five thousand years ago, Yoga is in recent times spreading all over the world. Still, its true meaning is but partially understood or misunderstood. Yoga is regarded as a form of exercise or acrobatics, or is considered a means of achieving physical health. It is also associated with mysticism and religious esotericism.

In reality, Yoga is a fundamentally religious combination of knowledge and self-discipline that speaks of truth. It recommends study, then experience, then meditation, then mastering the law of truth. To seek eternally the truth is the essence of Yoga and that which makes its character religious.

One truth is that the human being is simply a part of Nature, just one more living entity. As such, the human being is subject to the laws of Nature. Obedience to natural law is therefore the human being's only means of salvation or healing. At the same time, because according to natural law the human being's recuperative powers are innate and not extraneous, anyone can heal himself by himself. In fact, no one can be healed except by himself.

Both the philosophy and practice of Yoga aim at a balanced life-style in order that the individual may be enlightened and life enjoyed. This involves the processes of self-improvement, self-development, self-stabilization, and self-preservation. In other words, Yoga is a path of self-transformation, which is self-healing through individual effort. Because it teaches us to regain and increase the healing ability with which we were born, Yoga is the natural, independent way to normalcy or health.

Assuredly, Yoga is not a "how to" method of improving mental or physical imbalance. But it is a way of removing the causes of mental and physical imbalance, that normalcy may follow naturally. In this sense, normalcy equals a natural life-style which equals health. This equation is the way of Yoga.

Yoga addresses the whole life-style. It emphasizes the body and the mind as well as the spirit. The cure of any disease, including defective vision by means of Yoga, necessitates changing the life-style. In so doing, you can not only achieve the tangible benefit of good health but can also discover who you are, thereby reaching a higher stage of personal human evolution.

Yoga and Eyesight Theories

The various medical and popular opinions concerning the cause of nearsightedness are disparate and inconclusively proved. For example, nearsightedness is often considered to be the result of habitual reading under insufficient light. A medical theory attributes it to a stiffening of the eye muscles, a condition which, gradually worsening, results in paralysis of the eye nerves.

Methods that have been put forth to cure the eyesight are likewise variegated. While some people do claim the recovery of their vision through a particular method, others who have applied the same method register failure. Instead of blindly accepting a theory on nearsightedness and blindly following a particular method of recovering it, the Yoga procedure is: 1) to note the given theories on eyesight deterioration; 2) to heed the known methods and examples of vision recovery; and 3) to use these as references to find one's own way.

Yoga and You

It is up to you to discover what you need to heal your vision, which is to discover yourself. You are advised to examine the common theories on nearsightedness and the methods purporting to improve it. But remember that all human beings are different—this is fact. And since there are myriad variations from individual to individual, no one should be expected to embrace any theory or method unconditionally. You are asked therefore to test the efficaciousness of any theory and method on and for yourself.

This book shall initially relate the personal experience of the man upon whose teaching it is based, Masahiro Oki: how for many years he suffered from nearsightedness and eventually cured himself of the affliction; and how he realized what qualities and habits he had, and to what extent he had them, in common with other nearsighted people. Please recognize the qualities and habits you have in common with what were his and be counseled by his experience, deriving clues from it for yourself.

You will be informed of the theories formulated by medical science and by Masahiro Oki to explain nearsightedness and the methods devised to rectify it. Please make use of what is appropriate for your own vision recovery. You will also learn of meditation and the corrective exercises Masahiro Oki designed for eyesight problems. While meditation is the optimum means of developing your natural healing power, corrective exercise enables you to find the cause of your sickness as well as its recovery.

Throughout the book, in various guises, you will be given the principles of self-healing according to natural law. Please refer to them to create your own way of healing yourself.

Chapter 1

The Experience of Nearsightedness

As related in biographical accounts found in several books published in Japanese by Masahiro Oki, he in his lifetime suffered serious diseases, impaired vision being but one of them. While yet a boy, he understood nothing about sickness. But once out of school and into the world, it did not take long for him to ascertain that his mental and physical condition shared much in common with that of other nearsighted people. It also became apparent that everyone else, just as he, was seeking freedom from suffering, from the pain of nearsightedness or some other equally troubling ailment.

Throughout his young years, Masahiro Oki tried a variety of cures for poor vision and for every other sickness that arose within him, but none was successful. He admitted that he was ignorant of natural law. Not until much later, upon reflection, did he recognize how wrong his many attempts had been. He had misunderstood the cause of his every illness, including nearsightedness. It was only through the experience of regaining his health entirely, especially through the experience of curing himself completely of myopia and then of cancer, that he was able to discover there is a way to heal disease. Whatever the illness, the principles of healing are the same.

No disease can be recovered without eliminating its cause and without following natural law. This is the lesson Masahiro Oki learned and found to be companionable with the widsom of Yoga. Freedom is unattainable without practicing a life directed toward *satori*, or enlightenment, which simply means to experience and then to know the truth. To know the truth experientially, to have awareness, is to be able to act upon it. This process, the fundamental emphasis of Yoga, is the process through which a person achieves enlightenment and is the same process through which a person heals himself.

Eyeglasses as a Recovery

For your consideration, perhaps to compare with your own circumstances, here is Masahiro Oki's account of the onset of his nearsightedness in childhood:

During the fourth year of elementary school, my eyes began to bother me, frequently hurting. I was prone to headaches, the lessons on the blackboard became unclear, and when reading I had to hold the book close to my face. The vision problem was particularly aggravated when studying under little light.

On the order of several teachers, I was examined by the school physician who diagnosed the problem as myopia and prescribed eyeglasses. Although some of the other children also squinted when trying to see the letters on the blackboard, no other child in the Korean countryside where I grew up was given glasses; yet some of those children must have had eyesight problems too, though perhaps not so severe as mine. Nevertheless, there I was—one of the few Japanese children in that part of Korea and the only youngster wearing glasses—a "cure" agreed upon by the doctor, the teachers and my parents.

The adults around Masahiro Oki were never able to explain correctly the cause of his affliction. The school physician, in accordance with accepted medical opinion, believed that nearsightedness is hereditary. Not until much later did Oki realize how little heredity has to do with nearsightedness and how much life-style does. After leaving home to travel through Asia, he noticed a prevalence of myopic people in urban centers as opposed to scant few in the countryside. He also observed an absence of nearsightedness in countries technologically undeveloped. Although Oki later became certain that heredity is rarely responsible for myopia, in childhood he believed the school physician's diagnosis that his poor eyesight was inherited.

He went on to say:

The physician's advice was to wear glasses constantly, to read only under bright light, and to eat nutritious food. But alas! —neither he nor any of my well-meaning advisers understood the truth behind my condition. Every adult known to me held the superstitious-like attitude that absolutely the only solution to the problem of myopia is forever to wear glasses. Although I followed the physician's recommendations as best I could, my eyesight did not improve but continued to deteriorate.

Wearing glasses will never improve myopia. It will in fact weaken the eyes further. It is better, if you must wear glasses, to wear them intermittently. Taking off the glasses is a basic step toward improving the condition. But in the case of Masahiro Oki in childhood, he thought, because he was told so, that removing his glasses would increasingly hurt his eyesight. This was an especial dilemma for him growing up in Korea:

At that time the Korean culture demanded deep respect for one's elders. It was considered impolite, even insolent, to wear glasses in front of a superior. Since I lived in an area where tradition was stringently preserved, I had to

> obey. Whenever confronted by an adult, except for my parents and most of my teachers, I would quickly remove my glasses. Whenever I put them on again, I would feel dizzy. Mistaking the dizziness as a necessary part of my sickness, I withstood it with patience, and whenever custom allowed I kept my glasses firmly on.

The dizziness, headache and inability to focus well that often accompany taking off or putting on one's glasses are actually signs that the eyesight is attempting to correct itself. These discomforts are signals of adjustment. They are messages saying that the prescription of the eyeglasses is presently inappropriate. In other words, the body is reacting against the eyeglasses.

Eyesight is inconsistent. Glasses are not. Glasses lock the vision into a particular corrective code, disallowing change. Yet it is normal for the vision to change, including to change for the better. This is especially so at times of overall improvement in one's health or heightening of one's healing power. There were periods in Masahiro Oki's childhood when he could see better with his glasses off than on, but he never knew why. If a person is ignorant of the meaning of visual changes or misreads signs of adjustment, as Oki did as a youngster, the eyesight cannot be improved. To improve the eyesight it is necessary to understand vision theory correctly, to apply methods right for you, to recognize signs of improvement, and all the while to develop your own natural healing power.

Improvement Naturally

Masahiro Oki's elementary school education was supported by a home atmosphere of spiritual discipline. His parents were religious people. On various evenings his mother would invite a monk or priest to conduct lecture/discussions at their home for which neighbors would gather, bringing food. Also, each Sunday the family would fast and friends would convene to practice Zazen (sitting meditation) with them. As he said:

> With every Sunday of fasting and Zazen my vision improved. Each Monday morning when I put on my glasses for school, dizziness and pain resulted. Unfortunately, I did not then know that mental tension, leading to physical tension, is the major cause of nearsightedness and that the fasting and meditation had relieved these tensions. When a person is already tense or overexcited, it is easy to miss or overlook important signs of healing. Thus I could not correlate the cause and effect syndrome operating within my own case of natural improvement.

When the mind is under stress the body tenses too, paralyzing the nervous system. The eye nerves and eye muscles are of course subject to the paralysis.

Relaxation is essential to the natural healing process and two keys to mental and physical relaxation are Zazen meditation and fasting.

There was another childhood experience that alerted Oki to an aspect of relaxation when he recalled it years later:

> For the elementary school class in reading-and-writing Chinese characters, we students were required to practice aloud, reciting from the blackboard. The result was a chanting kind of vocalization, automatically supported by long, deep breathing, that caused our bodies to sway from side to side. Also, since the teacher of that particular class forbad me to wear glasses in front of him, it was left for me to discern the characters on the blackboard by my own power. I soon found that I could read from afar quite easily without my glasses, and could continue to see well without the aid of glasses for several hours after each of these classes. The principle then at work and then unknown to me is that rhythmical breathing accompanied by body motion helps the vision.

Reading aloud is actually a breathing exercise. It enables the inhalation to deepen and the exhalation to lengthen, which relaxes the body. It also allows the rhythm of breathing to stabilize, which calms the mind. On the other hand, unrhythmical breathing, including breathing in spurts, stopping and starting the breath, or tending to hold the breath, causes the mind to become anxious and the eyes to fixate. Normal eyes, the product of mental and physical relaxation, are constantly in motion.

Another way to relax the body is to twist, stretch and shake it. You can verify the validity of this principle by recalling what happens when riding on a train. Due to the rocking motion, you grow sleepy. Whenever the body is shaken or let loose, it can relax. But when contracted or kept in a rather fixed sedentary state, the body as well as the mind become tense.

Breathing deeply and rhythmically, shaking the body, and intentionally moving the eyes, all serve to improve the eyesight. Although Oki benefited from the result of these principles in the reading-and-writing class of elementary school, he could neither articulate nor confirm them until undergoing training at a lamasery in Mongolia nearly twenty years later.

If you recall, the second recommendation of Oki's school physician, after the order to wear glasses constantly, was to read only under bright light. This he followed implicitly. But most bright light, he has reported, especially sunlight, only reflected off the white paper into his eyes, causing them to tire quickly.

If the cause of nearsightedness were reading in dimly lit places, then there would have been an epidemic of myopia before the discovery of electricity. But in all pre-modern eras, even in the days of reading by oil lamp, myopia was a rare disease. The problem with reading in dimly lit places is not the darkness of the place itself, but that the reader tends to strain his eyes toward the print and hunch his body over the book in order to catch the words. The resultant tension, if it persists in the posture and muscles of the eyes, is capable of creating nearsightedness.

It took time, but Masahiro Oki eventually discovered that the onset of his own nearsightedness was poor posture:

> I finally traced its beginnings to the second grade of elementary school when I sustained injuries to my arms and legs. As a result, I had taken to "protecting" those hurt parts by curving my spine and holding it tight. It was as if I were trying to ward off further injuries and compensate for those already received. So the tension began, and it was not until junior high school that I even found out about the poorness of my posture.

Life-Style Improvements

The hallmark of the Korean junior high school where Oki roomed and studied was Spartan education, somewhat in the style of army training. As he recollected:

> If a freshman's posture was sloppy or if he put his hands in his pockets, he would be hit immediately by a senior student. The attitude toward disciplining oneself was also expected to be severe.
>
> In the beginning, all this was hard for me. But I began to attend to my posture, rather than be hit, and in time allowed my spine to stretch up and open out of its distortion. By the time that the new, better posture had become second nature to me, my entire well-being had been favorably influenced. Also, my eyesight had improved.

Confessing a possible hypersensitivity, Oki explained he was generally unable to wear glasses that had become too strong for him. Each time his vision improved, he asked to have the prescription of his eyeglasses adjusted. Although this was admittedly helpful to his eyes, it was not the reason for their improvement. Nor was the improvement due only to the change in posture, though that was a telling factor.

A change in diet was another factor:

> At the junior high school dormitory I was able to fulfill completely the elementary school physician's third prescription which was to eat nutritious food. Since childhood I had been accustomed to a balanced, rather European-style diet. But the dormitory provided simpler meals, consisting mainly of whole grains, vegetables and seaweed. Meat, poultry and even fish were hardly ever served. This traditional way of eating soon proved beneficial to my eyesight.

The positive relation between simple, whole-foods nutrition and the healing process was confirmed to Oki after graduating college in Japan. From there he went to study at a Mongolian lamasery, where a Zen-style cooking was practiced. The lamasery fare, even humbler than that of the junior high school dormitory, was usually a bowl of brown rice or other whole grain along with soup and a vegetable side dish once a day. From the lamasery he traveled to a Yoga ashram in India for further study. There the diet was wholly vegetarian. By the time he left the ashram, his general condition and eyesight had improved remarkably.

Oki later thoroughly studied diet and nutrition in relation to healing and eventually taught that the kitchen is the best pharmacy. At the Oki Yoga Dojo (physical, mental and spiritual training center) he founded in Japan, simple macrobiotic food* has always been served in small quantities. The dojo way of eating has tested out as being a truly significant part of the dojo residents' improving their mental and physical ailments, including myopia. It is also a way of eating contributory to their participating assiduously in rigorous training from sunrise to sunset.

The good effects of Oki's junior high school changes in diet and posture were compounded by compulsory exercise:

> Most of the students were required to practice one or more martial arts plus various sports. Since I had been accustomed to martial arts since childhood, it was possible to enter into a vigorous pursuit of kendo (fencing), judo and archery. I also ran several miles everyday. Routinely I would remove my glasses for these highly physical activities. Through the exercise and sweating my eyesight naturally changed for the better after each session. My whole body became more flexible and muscles more resilient through increased active use.

Without exercise, the muscles of the body shorten and stiffen. This is also true of the muscles of the eyes. When the eyes are habitually kept relatively immobile, the eye muscles contract and become tight. The eyes dry up and inflammation results. There is then an urgent need for an increase in blood circulation to purify the eyes. Whole-body training stimulates the circulation, including circulation to the eyes, and makes the muscles fluent. Also, muscles need moisture to relax. The sweating that accompanies hearty exercise helps the muscles of the body relax, just as tears or moisture help the eyes relax. Whole-body training, especially when assisted by specific eye exercises, contributes immensely to improving the vision.

Behind every disease, and nearsightedness is no exception, there is some degree of nerve imbalance and muscle contraction, followed by inflammation, blood

* Macrobiotic food consists of whole grains, especially brown rice, as the staple, accompanied by in-season vegetables, sea vegetables, beans, soup and in-season fruit. It also includes the use of fermented soybean products known as *miso*, *tamari* and *tofu*. All foods are simply and cleanly prepared for their own food values, retaining their natural colors and flavors. For further information on macrobiotics and macrobiotic food preparation, consult the Suggested Reading.

impurity, and a decrease in circulatory power. In general, where there is sickness there is overly acidic blood and poor circulation. The origin of an acidic condition can be muscular tension, or faulty diet, or both.

To correct the acid/alkaline balance of the blood, improving circulation and assuring cleanliness, it is advisable to eat small quantities of simple food* and to engage daily in physical exertion to the point of sweating. The most effective means of alkalizing or balancing or purifying the blood is fasting plus exercise. With this combination most people can free themselves of elementary to intermediate myopia, anemia and other diseases inflicting the blood.

Another part of Oki's junior high school education was to sit in Zazen meditation as a means of improving the posture, breathing and control of the mind:

> Although most of the students resisted this discipline, I welcomed it. It recalled to me my father's example during childhood. In the beginning, however, I did find it extraordinarily difficult to breathe deeply, suddenly consuming complete intakes of oxygen. But in time I was able to accept more and more air. That too helped my eyesight.

Oxygen is the nervous system's primary nutrient and the blood's primary purifier. An increased consumption of oxygen is basic to healing any disease and meditation is one way to deepen the intake of oxygen.

A final requirement imposed by Oki's junior high school was to take a cold bath daily, year round. About this he said:

> Although at the time I was unaware that the nervous system is stimulated and the sense organs activated by cold, I was aware of a refreshing feeling in my eyes each evening after bathing.

At the Oki Yoga Dojo in Japan, almost everyone, regardless of age, is encouraged to take a hot bath followed by a cold one at the end of each day. This practice, along with the trainings in posture correction, deep breathing, physical exertion, mental relaxation and simple nutrition, cannot help but revive the participant's healing power. Through the dojo training, which is a process for recovering health naturally, the acuity of all five senses can be markedly restored.

Oki acknowledged that the life-style demanded by his junior high school mended his vision radically, and since incorporated valuable aspects of that life-style into the Oki Yoga Dojo training. But while yet a schoolboy, he remained ignorant of the theories behind his own experiences, ignorant of why his eyes, poor since childhood, could see during adolescence better than before. In other words, he remained unconscious of natural law and the way of healing.

---

* Simple food consists of whole rather than partial food and natural rather than refined or chemically processed food. In this sense, whole grains, sea and land vegetables, and small fishes (which can be eaten whole) are whole foods. Meat and animal products are partial foods. Refined flours and sugar are of course unnatural foods.

Regression

Near the end of Masahiro Oki's high school studies, he was inadvertently dealt a blow during a kendo match that resulted in pleurisy. Due to incompetent treatment, the pleurisy degenerated into tuberculosis and his eyesight became weaker than ever. He then thought the tuberculosis must surely have been the cause, but later recognized his diagnosis was wrong:

> The actual cause of my decline in vision was the poor posture and shallow breathing that go hand-in-hand with tuberculosis. I had assumed a stooping posture, hunching the upper back in order to protect my infected lungs. The posture of course caused blood to congest in my shoulders and neck. The congestion caused a decrease in circulation to the eyes and an increase in interocular pressure. Automatically, my eyes became tense. Since exercise was prohibited and rest prescribed by the physician, that tubercular period of "easy living" halted the improvement of my eyesight and sent it into deterioration.
>
> Upon my return to the high school dormitory, the vision problem was exacerbated. Even though it was the term of study for college entrance exams, the electric lights, according to rule, were turned off at 10: 00 P.M. The graduating seniors were compelled to read by flashlight in the attic, in closets or in bed. In a short time many students exhibited symptoms of myopia.

The obvious and conventional deduction from this account would be to blame studying in dimly lit places for the students' collectively impaired vision. But that was not the case. The actual reason for the emergence of myopic symptoms was twofold: 1) the stooping posture they were forced to assume in order to read in confined places under limited light; and 2) the hypertension produced in anticipation of the examinations. The cause of their nearsightedness was mental and physical tension.

The way to improve nearsightedness, or any disease, is to create mental and physical conditions opposite to those which produced the disease in the first place. A calm, peaceful mind and a relaxed body allow the eyes to see clearly.

Progression

From Oki's years of college in Japan on into adulthood he was involved in a variety of activities in sundry places:

> With each change of job or location, my vision either improved or lapsed. Accordingly, rather than attend to my vision properly, I had the prescription

> of my glasses adjusted. In other words, once out of high school, wearing eyeglasses became a fact of life.
>
> One of the most notable changes in my ability to see occured while working in espionage just before World War II. My assignment was to infiltrate Tibet, but prior to that, as preparation, I was sent to Mongolia to study Lamaism that I might impersonate a lamaic monk.
>
> While in residence at the lamasery, I experienced what seemed to be a miraculous improvement in vision. In fact, the huge, endlessly stretching plains of Mongolia, which compel the inhabitants to gaze unintentionally and frequently at the evervisible horizon, were responsible for the "miracle." This gazing into the far distance followed by focusing on sights near at hand, as well as viewing the beautiful sunrises and sunsets across the plains, exercises naturally the eyes of the Mongol people.

Distance-gazing stretches the muscles supporting the eyeballs so that tension is released. Conversely, near-looking contracts and tenses the eyeball muscles. To maintain eye muscle elasticity and balance, it is necessary to use the eyes for looking alternately near and far.

Active engagement in the daily life of the lamasery aided immeasurably in Oki's eventual complete recovery of his eyesight. The recitation of sutras and chanting of mantras, the purpose of which was to develop concentration and will power, were contributory. He was also taught how the eyes function and was drilled in techniques to develop each of the five senses. He has since designated the lamaic temple training, because of its being both physical and spiritual, a training for life; that is, a training that weds practices in *dhyana* (attainment of the highest stability) with practices in *samadhi* (attainment of unification with the object of attention, the immediate action, the surroundings).

He said:

> It was a privilege for me to stay at the lamasery and I am grateful for the intense trainings given me that later influenced my own Yoga teachings. I was introduced to Lamaism, a Buddhist discipline originally brought to Tibet by Indian Yogis. Lamaism stresses *dharana* (Yogic concentration). I was also exposed for the first time to yin/yang theory and its practical applications. Even though this eminent Chinese philosophy known as Taoism had been brought to Japan by the monk Kobo-Daishi in the ninth century and was later incorporated into the Shingon sect of Buddhism, it was new to me. Also, the *dhyana* (stability) practices taught me were none other than fruits of Zen Buddhism. I am therefore indebted to the lamasery for revealing to me the philosophy and practices of Indian Yoga, Chinese Taoism and Zen.

The teachings of these three disciplines finally led Masahiro Oki to apprehend the human being's tremendous capacity for seeing, for understanding natural law, for self-healing, and for self-realization.

Chapter 2

The Treatment of Nearsightedness

The earlier a disease is discovered, the more easily it can be improved. Cancer, for example, is particularly stubborn because the symptons do not display themselves until the disease is well established. Myopia, however, is almost immediately detectable due to its major symptom, an inability to see as clearly as before. Practically speaking, insofar as the correct theory regarding the cause of nearsightedness is pinpointed and insofar as the correct method to adjust the vision is undertaken, nearsightedness can be readily discharged when yet in an elemental stage. More advanced myopia takes time.

The following are symptoms of nearsightedness or various ways to detect myopia early on:

- Can no longer read or study for an extended period.
- Can no longer read a book held more than nine inches from the eyes.
- Begin to watch television from a closer position.
- School marks or work record deteriorates.
- Mentality and action suddenly become negative.
- Become irritable and lose good communication with others.
- Become less impressed by beauty than used to be.
- Have headaches and heaviness of head; feel tired and tense in the shoulders.
- Posture begins to stoop.
- Begin to squint habitually.

And finally, even though myopia is not hereditary in almost all cases, if your parents are nearsighted, the probability of you also becoming so is greater than if your parents see well.

Anatomy of the Eyes

The eye is made up of the eyeball and the optic nerve which extends from the rear of the eyeball to the brain. The eyeball itself is composed of various parts that may be compared to the parts of a camera. As shown in Diagram 1, the

Diagram 1

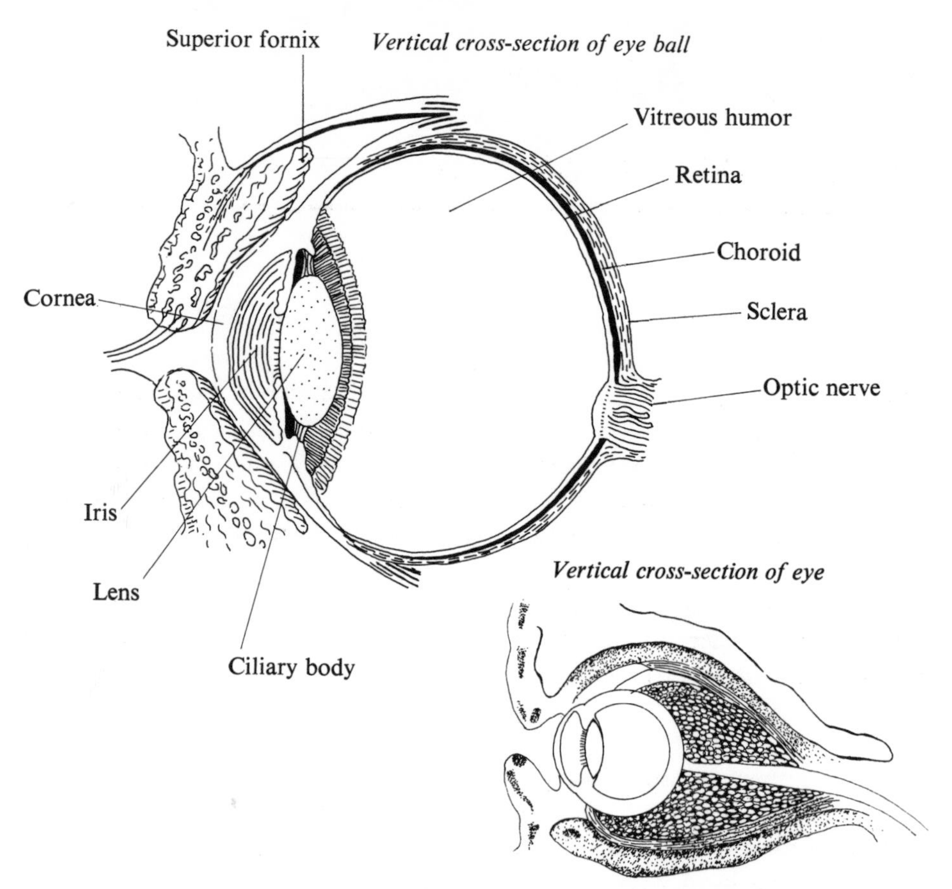

sclera which is a fibrous membrane forming the white or outer coating of the eyeball that at the very front becomes the almost perfectly transparent tissue called the *cornea*, is as the body of a camera. The *iris*, which is the pigmented membrane surrounding the *pupil*, equipped with muscles to adjust the size of the pupil for regulating the amount of light entering the eye, is as the diaphragm of a camera. The *crystalline lens*, which is a transparent tissue situated just behind the iris for the purpose of refracting or bending light admitted by the pupil, is as the lens of a camera. The *retina*, which is the innermost lining of the eyeball, actually a membrane extension of the optic nerve upon which the entering light or image falls, is as the film of a camera.

Though similar in parts, the eye and the camera are not operatively the same. The camera is a machine with independent, replaceable parts; while the eye is a living organism with interdependent parts which needs a connection to the brain and a supply of blood for it to operate. The *choroid*, which is the middle lining of the eyeball that in front becomes the iris, is composed largely of blood vessels and so feeds the eye. The focusing of the camera and the eye also differs. A camera lens cannot adjust itself to bring into focus the object to be photographed; the distance between the lens and the film must be adjusted by the photographer.

But the eye can bring objects into focus by itself, by virtue of the *ciliary muscle* which adjusts the shape of the crystalline lens. This muscle, admitted to be the chief agent in accommodation, that is, in adjusting the eye to the sight of near objects, consists of two sets of fibers, radiating or longitudinal and circular or ring-shaped. Together these function to change the shape of the crystalline lens from thinner to thicker, that is, from less to more convex. As the shape of the lens changes, it is able to focus light rays from far to near, making an image on the retina. The image focused on the retina, as on the film of a camera, is inverted and reversed. But since the retina envelops all the optic nerve endings that merge into the brain, the mind is able to interpret the image so that no confusion arises. In this way, according to accepted physiology, the eye sees.

Accessory to the eyeball are the eyebrow, the eyelid and eyelashes, the conjunctiva, the lachrymal gland, and six extrinsic muscles. These accessories either protect the eye or help it function. The conjunctiva, for example, is a mucous membrane under the surface of the eyelid that converges over the fore part of the sclera and the cornea. From the conjunctiva a liquid is secreted to aid in both eyelid and eyeball mobility. The lachrymal gland, located at the upper outer angle of the eyeball, secretes tears. Tears help the eyeball move smoothly, remove dust from its surface, sterilize the eye, and allow for blinking. Besides the internal eye muscles controlling the pupil or the crystalline lens, there are six muscles outside the eyeball for holding it in place and regulating its directional movements. These extrinsic muscles, as shown in Diagram 2, are so arranged and innervated that both eyes move together rather than independently.

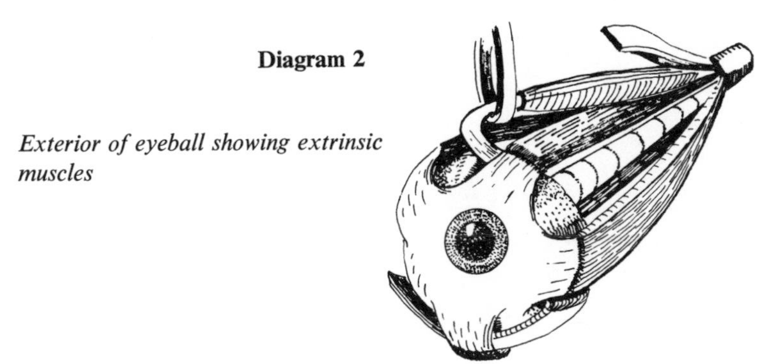

Diagram 2

Exterior of eyeball showing extrinsic muscles

Three Kinds of Myopia

The first kind of myopia is a transient state better referred to as pseudomyopia. It is due to tension in the ciliary muscle, generally arising from a tendency to stare at objects nearby for extended periods without looking into the distance. As a result, the circular fibers of the ciliary muscle, contracted to thicken the lens for near-looking, become tight; while the radiating fibers weaken from insufficient use. Pseudomyopia exists only when the ciliary muscle is tired and is therefore a temporary condition. It can be relived by intentionally looking into the distance, that is, intentionally alternating one's focus from near to far, so as to use harmoniously the fibers of the ciliary muscle. Temporary far sight can be had by the application of atropine, a chemical derivative of the deadly night-shade *Atropa belladonna*, which apparently paralyzes the ciliary muscle and so prevents near-sight accommodation. Nevertheless, it is preferable to relax naturally the fibers of the ciliary muscle through balanced use, lest pseudomyopia turn into actual myopia.

The second kind of myopia, on the road to actual myopia, is a form of sclerosis. It is characterized by persistent tension in the ciliary muscle, whereby its circular fibers become rudimentary or even absent. The ciliary muscle cannot then thin the shape of the lens sufficiently to bring far objects into good focus.

The third kind of myopia, a result of chronic sclerosis, is actual myopia. When the lens is held in a near-permanent state of convexity, engorgement or blood congestion arises within the eyeball. To compensate for the lens' inability to change shape, the eyeball itself changes, becoming an ellipse as shown in Diagram 3. An elliptical eyeball is responsible for sustained error of refraction. In the case of myopia, this means that the parallel light rays from distant objects, though not

Diagram 3

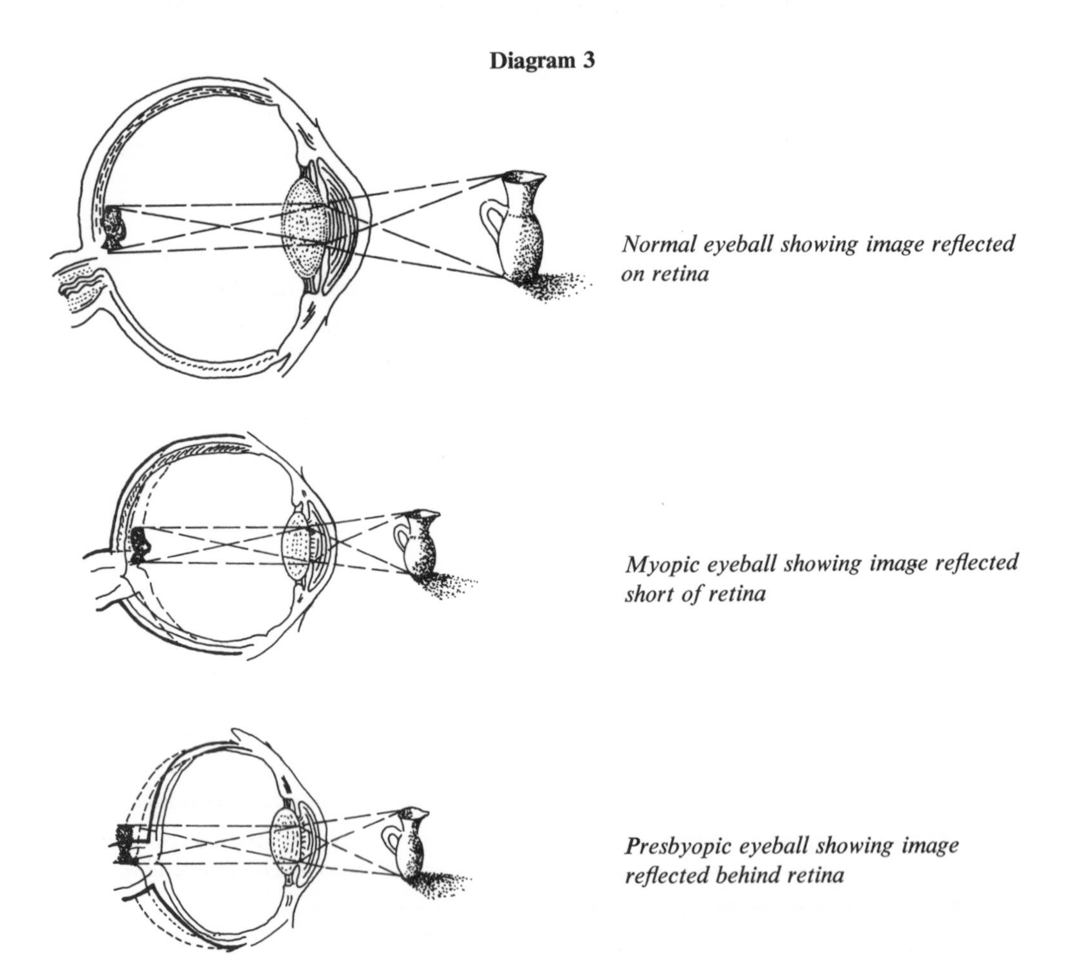

Normal eyeball showing image reflected on retina

Myopic eyeball showing image reflected short of retina

Presbyopic eyeball showing image reflected behind retina

the divergent light rays from near objects, are focused at all times short of the retina—because the eyeball itself has become too long.

Insofar as engorgement worsens, that is, insofar as the eyeball is allowed to remain elliptical or become more so, the quality of the retina might also begin to change. Distance might accrue between the retina and the blood vessels of the choroid feeding it. If this distance becomes too great due to increased engorgement, the retina can actually detach from its bed in the eyeball, usually causing total blindness.

Myopia and Presbyopia

Another condition in which the ciliary muscle as well as lens flexibility are afflicted is presbyopia, a difficulty in seeing nearby objects. To see near, the lens must be thickened from its normally flat state by power of the ciliary muscle. But in presbyopia, due to weakness of the ciliary muscle, accommodation ceases. When sclerosis sets in, the eyeball too is affected, shortening in shape. Thus objects of near vision are focused beyond the retina, as shown in Diagram 3.

Aging of the ciliary muscle is the accepted cause of presbyopia. Young people can clearly see an object placed as near as three inches to the eyes. But as the body matures, the capacity to see nearby decreases so that an object must be thirteen or more inches from the eyes in order to see it clearly. This is a symptom of senility. In general, symptoms of senility appear at about the age of forty. But in the case of the eyes, since by the age of twelve the eyeball is already fully developed to adult size and stops growing, senile symptoms might manifest themselves earlier. Even by the age of ten there can be some loss of lens flexibility or accommodation. If a person is sick in his youth or exhibits signs of aging early on, it is likely that presbyopia will be present before the age of forty, though for most people presbyopia becomes acutely apparent at sometime between the ages of forty and forty-five.

Needing to hold a newspaper or book farther from the eyes than is customary is the first symptom of presbyopia. Another is that the eyes tire easily when reading and that a headache, even double vision, erupt. Or it might be that when initially reading the eyes see well but after a while the vision becomes hazy. To compensate, a presbyopic person leans back from a book or newspaper in order to increase the distance between the object and the crystalline lens. A final symptom of presbyopia is an ability to see well in the daytime but not so well at night. This is because the pupil automatically becomes smaller when exposed to light. And since it is normal for the pupil to slightly contract during the process of accommodation, the crystalline lens does not need to achieve a complete thickness in order to see near in daylight. The same effect is generated by myopic people through squinting; the pupil is contracted intentionally in order to see objects fairly near at hand more clearly.

A myopic person may or may not be presbyopic as well. The two are distinguishable according to the symptoms outlined, and neither should be confused with "farsightedness" which is better called hypermetropia. In fact, what is usually called farsightedness is not farsightedness at all. Nor is it the opposite of nearsightedness. Hypermetropia is rather an inability to see clearly at a near point or at a distance. In this sense it is a more serious disease than myopia. People over the age of forty who can see well neither near nor far, rather than being hypermetropic, are likely suffering from a combination of myopia and presbyopia.

The initial characteristic then of both myopia and presbyopia is tension and weakening of the ciliary muscle. A simple way to rectify both these conditions early on is to exercise and relax the ciliary muscle, indeed to relax the whole upper body. Balanced vision depends in great part upon relaxation of the eye muscles, both intrinsic and extrinsic, so that the parts of the eye might move freely and the eyeball itself might move continuously. Normal eyes are in motion, while eyes that seem constantly to be still are abnormal and tired. Inactive eyes indicate a fundamental imbalance in the nervous system, which may be stated as a condition of habitual mental and physical tension.

Although the need to relax the eyes has become chronic and widespread in modern times, it is not a new phenomenon. According to recorded history, there were eye doctors practicing in ancient Egypt and India. In fact, ten percent of all medicine in the earliest civilizations was concerned with eye treatment. In both Egypt and India people's eyes, taxed by the ever brightly shining sun and plagued by numerous tiny flying insects, were prone to disease; hence the need for eye medicine. As a palliative, ammonium sulfide, a black powder, was applied to the eyes to cleanse and relax them. In time, the powder was put under the eyelid as a prophylactic to eye infection, and since it made the eyes appear longer, women began to use it cosmetically. There is also at least one example of a device for eye protection in Roman history. The Emperor Nero apparently wore round green lenses, perhaps the first instance of sunglasses, to shield his eyes. In later times in Japan, specifically during the Heian period or the ninth through twelfth centuries, people applied dried eels to their eyes. Eels are rich in vitamin A, especially beneficial to the vision, but beyond that fact, eels were considered a guard against night blindness. In addition, history records various herbal applications to relax the eyes and even the case of using children's urine as an eye medicine in the Orient.

The historical understanding of a connection between eye muscle relaxation and balanced vision is certainly applicable to modern times when so many people use their eyes incorrectly. Myopic people tend to stare, fixing their eyes on one point, a near point. A myope must consciously mobilize his eyes; that is, change his use of them. Change, a tenet of natural law, is that factor which allows for a balance to be struck between tension and relaxation.

While working to change the use of your eyes, you should also strive to balance the entire nervous system. Although medical science implicates the eye muscles as the chief antagonist to seeing well, Yoga teaches us that any part of the body must be reckoned within light of the whole. Eyesight and the ability to focus is not the province of the ciliary muscle only. Eyesight, since the eyeball is connected to the brain, is the province of the entire nervous system. Sight imbalance must be recognized as related to tension of the mind and body, especially the upper body, not just to tension of the eye muscles. It is extraneous mental and physical tension in general that makes and keeps the eye muscles tense.

If one were to insist that the muscles of the eyes are the sole culprits in causing vision problems, then one could easily conclude that such phenomena as poor light while reading or excessive television viewing are producers of myopia. This is the view of medical science. As for the cure of myopia, medical science admits none, recommending only the wearing of eyeglasses. But to wear glasses is to derive the eye muscles the opportunity ever of relaxing. Glasses are part of a vicious cycle that does not cure but promotes myopia. The only reasonable conclusion is that most medical authorities do not know either the fundamental cause of nearsightedness nor the natural way of healing it.

The only way to heal actual myopia is to change the life-style. Most people are dependent upon a number of set patterns, including one way of moving, one way of thinking, one way of eating, and one way of using the eyes. All these affect the vision, and any or all of these can create the mental and physical tension that causes myopia. If the diet, for example, is overly acidic or overly abundant, you can be sure that the acid/alkaline condition of the blood is unbalanced, even that the blood is dirty, and that the blood circulation, including that to eyes, is sluggish. Under such circumstances, the nervous system and the eyes cannot perform well. In order to adjust your vision, you must intentionally and consciously adjust your life-style—movement pattern, mentality, eating habits, use of your eyes—so as to balance the nervous system.

Eye Training

Eye exercises are one way to change the use of the eyes. When Masahiro Oki was residing at the lamasery in Mongolia, he was taught techniques to heighten the power of all five senses. The specific exercises given him for the eyes are methods to mobilize and relax them, sometimes through relaxing the mind and body. The eye exercises below are derived from Oki's lamasery training and Yoga practices.

Prior to performing the exercises, do five to ten minutes of general loosening-up movement. Take the time to enjoy each eye exercise for itself, not depending on a particular goal or result. Exhale deeply along with the given movement, letting your face and mind relax. Each exercise may be practiced two to three times. After each group of repetitions shake the upper body suddenly, quickly and fully, and then relax for a moment before proceeding to the next exercise. The entire series is best practiced daily, even twice a day, for up to twenty minutes a session.

For the following eye training exercises, kneel in the seiza posture with the hips

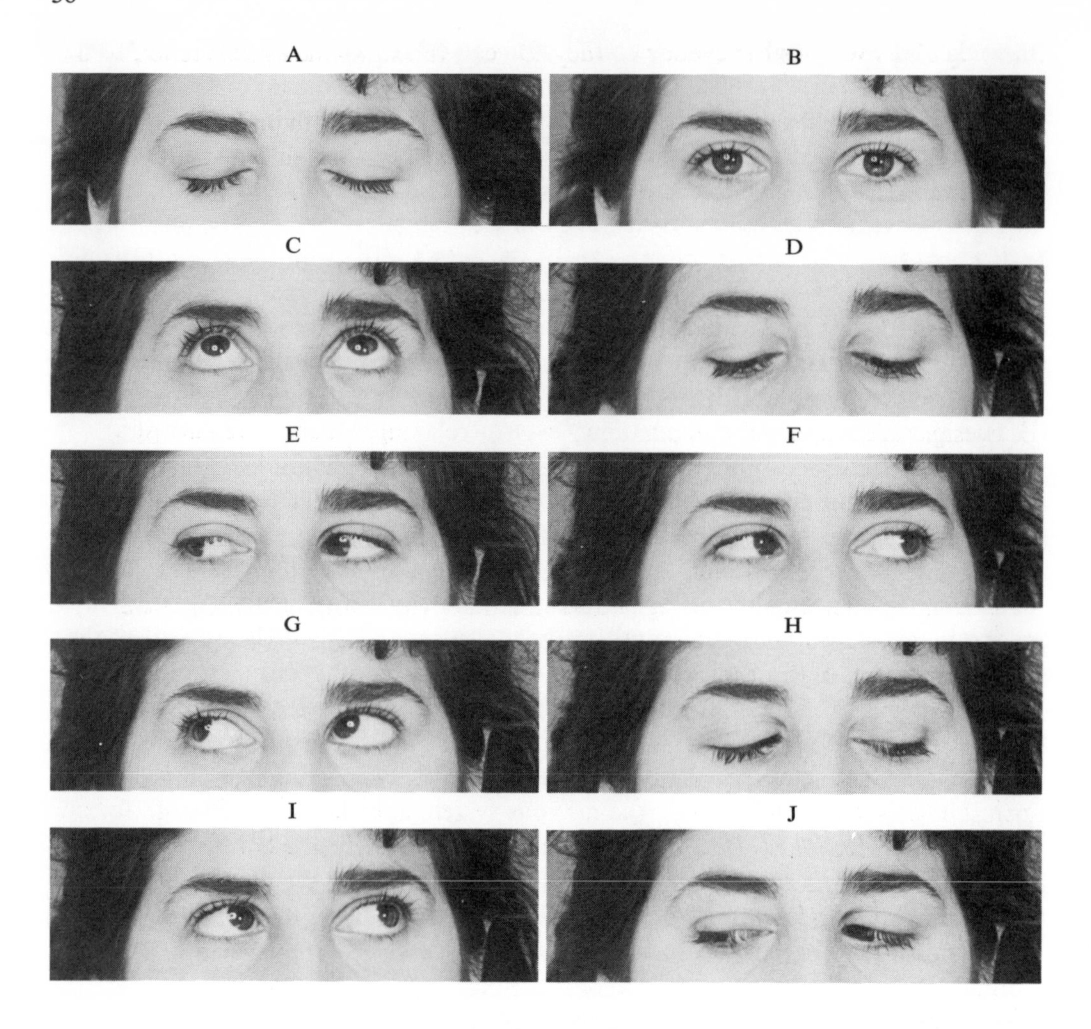

between the heels, or sit on the edge of a chair keeping the spine upright, or stand with the spine stretched. (These three postures are illustrated on pages 134 to 136)

1. Close your eyes. Inhale deeply. Exhaling, rotate your neck and head. Then rotate the shoulders both forward and backward. Next, close your eyes tightly and release them suddenly. Repeat several times. To increase the effect, especially if you do desk work (and perform this exercise intermittently while doing desk work), arch the spine and neck backward while closing tightly and releasing your eyes. You can apply the same technique to freeing tension from the shoulders. Inhaling, elevate and tighten the shoulders to the maximum and, exhaling, release them suddenly. Softening the shoulders also helps relax the eyes.

2. Blink repeatedly, fluttering the eyelids, until your eyes become moist. Or, exhaling, close your eyes tightly and, inhaling, open your eyes suddenly (A and B). Repeat several times and then massage your eyes. Tension in the pupil and muscles supporting the eyeball is a major source of tired eyes. When the eyes are moistened, they can relax. Blinking stimulates the lachrymal glands and hence a flow of tears. When consciously done, blinking can relieve, even prevent, muscle tension.

Blinking also exercises the eyelids so they too can relax. As the eyelids relax, so do the eyes.

3. Close your eyes and protrude them forward. Then pull them back. After repeating several times, look at an object. This exercise increases blood circulation to the brain while relaxing the eye muscles.

4. Focus on a point far in the distance for twenty to thirty seconds, breathing naturally. Maintain concentration on the point. Then blink your eyes rapidly several times. Next, focus on a point near at hand for ten to fifteen seconds, repeating the same procedure. This exercise relaxes the muscles controlling the pupil, and hastens their reaction to adjustments in distance.

5. With your eyes half-open, look up while exhaling. Since this exercise helps the eyesight focus itself, it can eliminate the tendency to squint.

6. Pull the chin straight back. Roll your eyes upward while exhaling (C). Hold for a few seconds and then, relaxing, return your eyes to center while inhaling. Roll your eyes downward while exhaling (D). Hold for a few seconds and then, relaxing, return your eyes to center. Repeat a few times.

Next, move your eyes to the right while exhaling (E). Inhale and return your eyes to center. Move your eyes to the left while exhaling (F). Inhale and return your eyes to center. Repeat a few times.

Then, look up at a forty-five degree angle to the right while exhaling (G). Inhale and return your eyes to center. Look down at a forty-five degree angle to the left while exhaling (H). Inhale and return your eyes to center. Repeat a few times.

Finally, look up at a forty-five degree angle to the left while exhaling (I). Inhale and return your eyes to center. Look down at a forty-five degree angle to the right while exhaling (J). Inhale and return your eyes to center. Repeat a few times.

At last, rotate your head clockwise a few times, rolling your eyes in a wide arc around your head while exhaling. Repeat counterclockwise. This exercise stimulates the extrinsic muscles supporting the eyeballs.

7. Pull the chin straight back and look ahead at an object. Exhaling, look from side to side several times without turning your head. Then, exhaling, turn your head from side to side several times, letting your eyes follow the head movement. Finally, exhaling, turn the whole upper body side to side several times, letting your eyes follow. Repeat the sequence with your eyes closed. This exercise also stimulates the exrinsic eye muscles.

After practicing the series of eye training exercises, do palm healing (as described below) on them, for optimum relaxation.

Eye Purification

The eyes can become tired not only from muscular tension but also from blood stagnation. In particular, if the eyes are used for an extended period while the body is kept fairly stationary, blood circulation to the eyes will slow down. The various external treatments offered below, some of which were taught to Masahiro Oki by the Mongolian lamas and others which belong to traditional medicine, may be applied to the eyes for stimulating blood circulation to them. In addition, it might be necessary to cleanse the entire bloodstream, perhaps by undertaking a period of fasting, so that circulation may become fluent throughout the body. Meanwhile, the following techniques should serve to refresh and purify the eyes.

Palm Healing: This technique is based on the understanding that *ki* or electromagnetic energy constantly courses through the entire body along identifiable meridians. One function of this energy is to induce and maintain vigorous blood circulation. Since the palms of the hands are a major locus of *ki*, that is, since *ki* is concentrated to a tremendous extent within the palms, you can stimulate blood circulation to any part of the body simply by placing your hands thereon. This principle is known intuitively and practiced by all human beings, as for example when we rub an ailing or tired part of the body or when we unconsciously rub our eyes if they have become tired or irritated. Palm healing, also called laying on of hands, is evidence of the body's desire and ability to heal itself by itself.

To practice palm healing, lie on your back or sit on the edge of a chair. Rub your palms together briskly for ten to fifteen seconds until heat is produced. Cupping your hands, place the centers of the palms over the eyes and close your eyes. Inhale deeply and while exhaling imagine energy or heat generating from your palms into your eyes. Hold this position for a few minutes while breathing deeply.

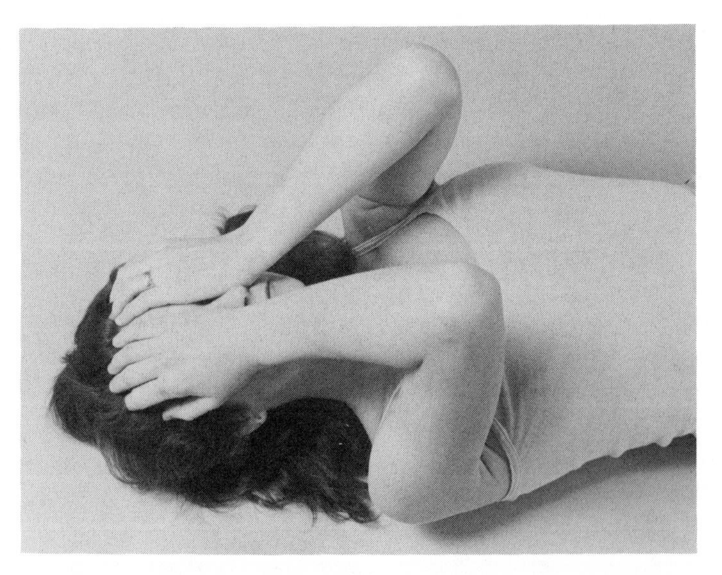

You can heighten the healing power of your palms by using them on yourself and others. It is advisable to practice palm healing on your eyes several times a day, especially for relaxation while reading or doing desk work and after performing the eye training exercises. Any closing of the eyes relaxes them. By adding the salient energy of your palms to closed eyes, you can bring circulation to them, release fatigue from them, and help clarify your vision by yourself.

Chlorophyl*: Chlorophyl, an extract of green plants that can be obtained in powder form, is a natural eye relaxant, far preferable to any synthetic preparation. It is mixed with distilled water to liquid consistency and may be applied daily as drops into the eyes. Masahiro Oki told of an occurance during his stay at the Mongol lamasery that introduced him to the benefits of chlorophyl:

> One day while out on the plains of Mongolia, a speck of sand blew into one of my eyes. The eye became inflamed and the swelling turned into conjunctivitis. As a remedy, the juice of a green plant similar to mugwort was squeezed directly into the injured eye by one of the lamas. Although it stung mightily for a few minutes, when the pain subsided I could see with absolute clarity. I thus learned that chlorophyl, since it contracts the blood vessels and strengthens the blood circulation within the eyes, is effective in improving the vision.

Although chlorophyl's spontaneous effect of producing what seems to be perfect vision lasts but two to three hours, its effect of stimulating blood circulation is long-range.

As an alternative to powdered chlorophyl or to extracting the juice from wild plants, you might try cucumber. Squeeze the juice from a six-inch slice of fresh cucumber. Add to this two tablespoons of whey and apply to your eyes, internally and externally. If whey is unavailable, the cucumber juice can be used neat. This lotion cools and soothes inflamed or tired eyes and tones up the eye membranes.

Herbs: While Masahiro Oki was in training for espionage, because he was to impersonate a medical practitioner amongst villagers in the Asian interior, he was provided with courses in Chinese and Korean herbalism. While using himself as a subject to study this form of medicine, his vision improved.

The following herbal remedies are traditional to Chinese medicine for the treatment of various eye disorders. They may possibly be obtained from a Chinese pharmacy, or alternative herbal eye treatments might be prescribed by an herbalist. Although herbs can be purchased in capsule form from a natural foods store, when prescribed they generally come in dried or powder form to be taken with water or brewed as a tea.

Ko-Ken-Tang (Pueraria combination)—to be taken at the onset of inflammation or swelling and pain in the eye. In cases complicated by constipation, add Chung-Huang-San (Cnidium and Rhubarb formula).

Yueh-Pi-Chia-Chu-Tang (Atractylodes combination)—to be taken for blepharitis (inflammation of the eyelid) especially when accompanied by swelling congestion, pain, secretion and lachrymation.

Hsiao-Ching-Lung-Tang (Blue Dragon combination)—to be taken for conjunctivitis (inflammation of the conjunctiva), irritation and lachrymation.

---

* If powdered chlorophyl is difficult to find in the United States, it can be purchased from the Oki Yoga Dojo in Japan under its Japanese name ヘリクロゲン (romanized as *herikurogen*). For inquiries, write to the Oki Yoga Health, Sawaji 777–1, Mishima-shi, Shizuoka-ken, Japan 411.

Hsi-Kan-Ming-Mu-San (Gardenia and Vitex formula)—to be taken for keratitis (inflammation of the cornea).

Pa-Wei-Ti-Huang-Wan (Rehmannia Eight formula)—to be taken for cataract, ache and chills in the waist area, and especially for weakness of the limbs.

Tzu-Sheng-Ming-Mu-Tang (Chrysanthemum combination)—to be taken for impotence, weakness, and eye disorders including asthenopia (eyestrain) and astigmatism.

Acupuncture and Shiatsu Massage: Another segment of Masahiro Oki's espionage training was instruction in both Oriental and Western medicine. During the term of these studies he received Chinese acupuncture treatments on his neck and the area surrounding his eyes, specifically for relieving myopia. Vision therapy by acupuncture is fundamentally a strong inducement for the upper body to relax. In particular, by inserting needles into key points along designated energy meridians, the shoulders, neck and eye muscles are made to relax while blood circulation to the eyes is increased.

Shiatsu is a form of massage based on the same principles as acupuncture, except that the thumbs and fingers rather than needles are used to provide stimulation. Also, it can be practiced on oneself.

As a preparation for shiatsu eye treatment, close your eyes and with the fingertips gradually press the eyelids inward and then push them from side to side. If you hear a liquid sound, there is excess mucus within the eyes. Next, with the heels of your hands press the temples inward and then rotate the temple areas ten to twenty times in both directions. Finally, with the fingertips press inward all around the skull bones encasing your eyes.

Referring to the Figures below, locate the eye treatment points on yourself. Each point should feel like an indentation, a hollow or a vacuum and will undoubtedly be sensitive for people with vision problems. The shiatsu finger-pressure technique is gradually to press in on an exhalation and release while inhaling. It is recommended to stimulate each of the following points three to four times.

1

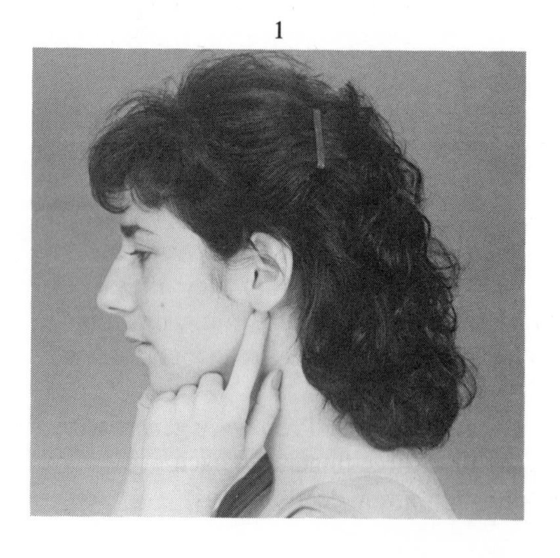

1. In the hollows just behind the junctures of the earlobes and the neck: Press with a thumb on each side simultaneously, pushing upward toward the eyes and holding through a few deep breaths.

2. Directly below the centers of the eyes, in the hollows of the cheekbones: Press with an index finger on each side simultaneously, pushing inward and then upward toward the eyes and holding through a few deep breaths.
3. On the temples: Press with a thumb on each temple, pushing inward and rubbing toward the eyes through a few deep breaths.

2

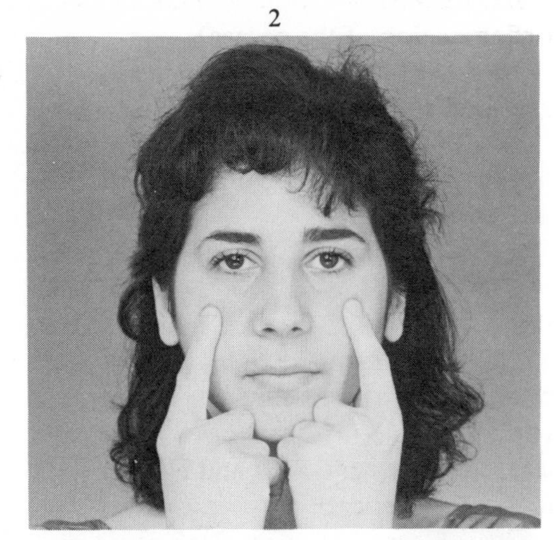

3

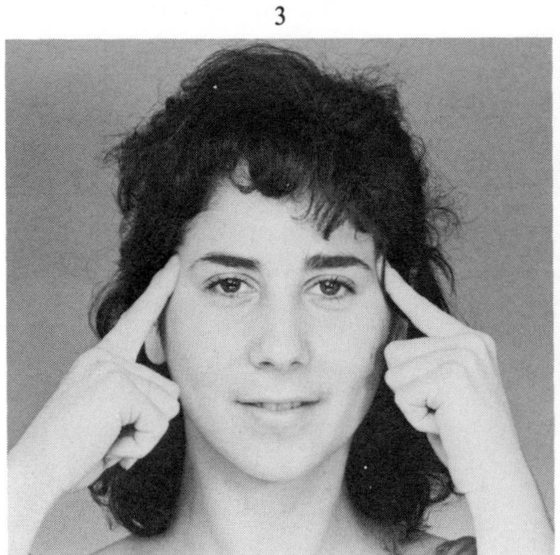

4

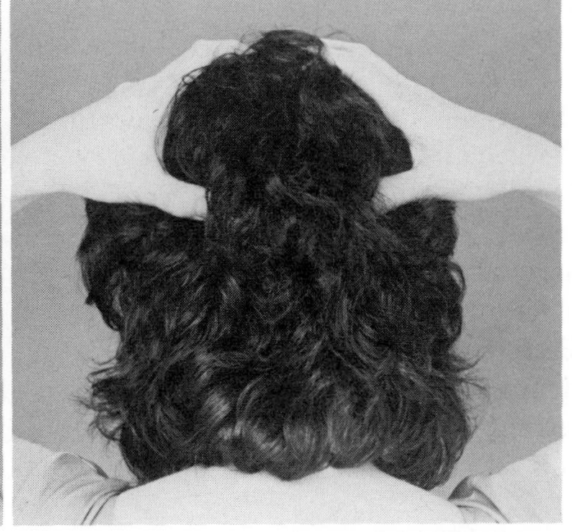

5

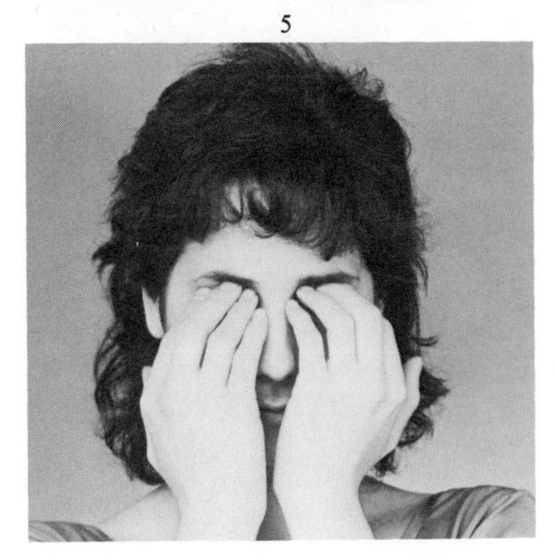

4. In the hollows on the back of the head directly behind the eyes: Press with a thumb on each point, pushing inward toward the eyes through a few deep breaths.
5. On the eyes: Press very gently with the first three fingers on each eye through a few deep breaths.

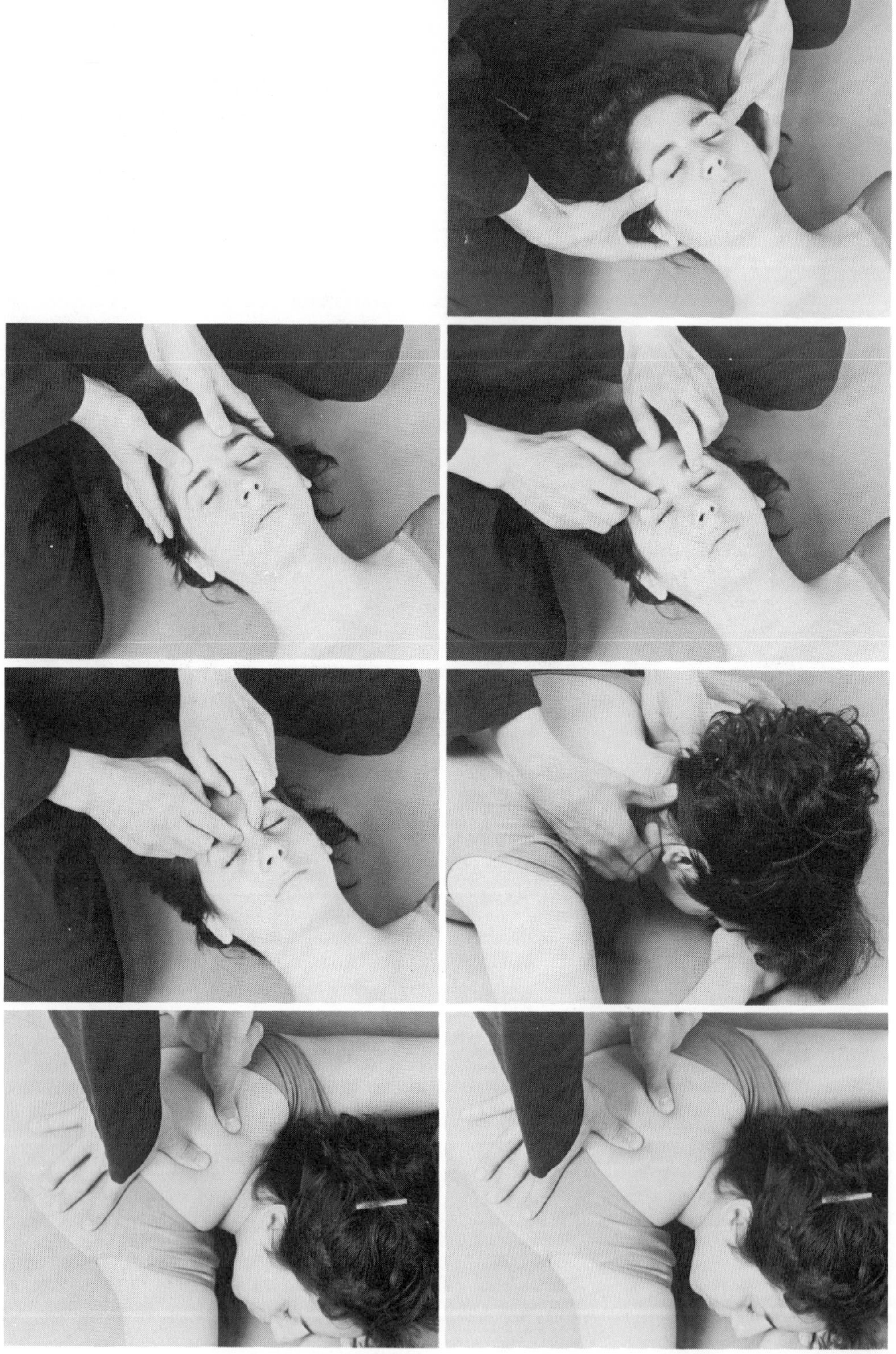

Chinese Eye Massage: In the Chinese tradition of health care there are several types of massage, including one for relaxing the eyes by bringing blood circulation to them. Although it does not pretend to cure myopia, Chinese eye massage is helpful in preventing further deterioration of the eyesight. Referring to the eye points shown in Diagram 4, practice the following massage techniques on yourself:

1. Place a thumb on that part of each eyeball just beneath the start of each eyebrow, and place your fingers on the forehead. With the thumbs, massage your eyeballs in a circular motion to a count of sixty-four times.
2. Place a thumb on that part of one eyeball nearest the nose, and place the index finger of the same hand on the corresponding part of the other eyeball. Push your eyeballs up and down to a count of sixty-four times.

Diagram 4

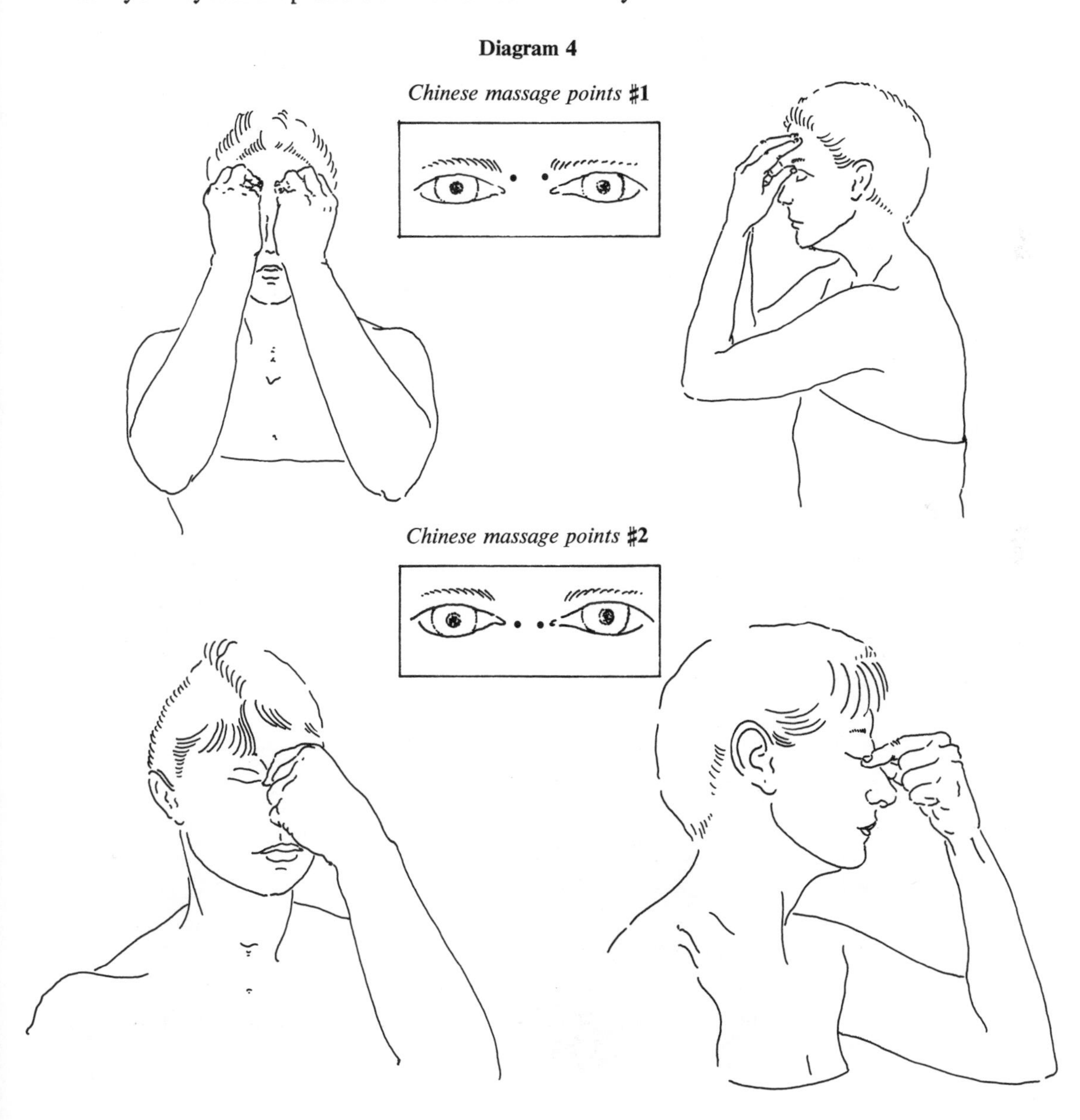

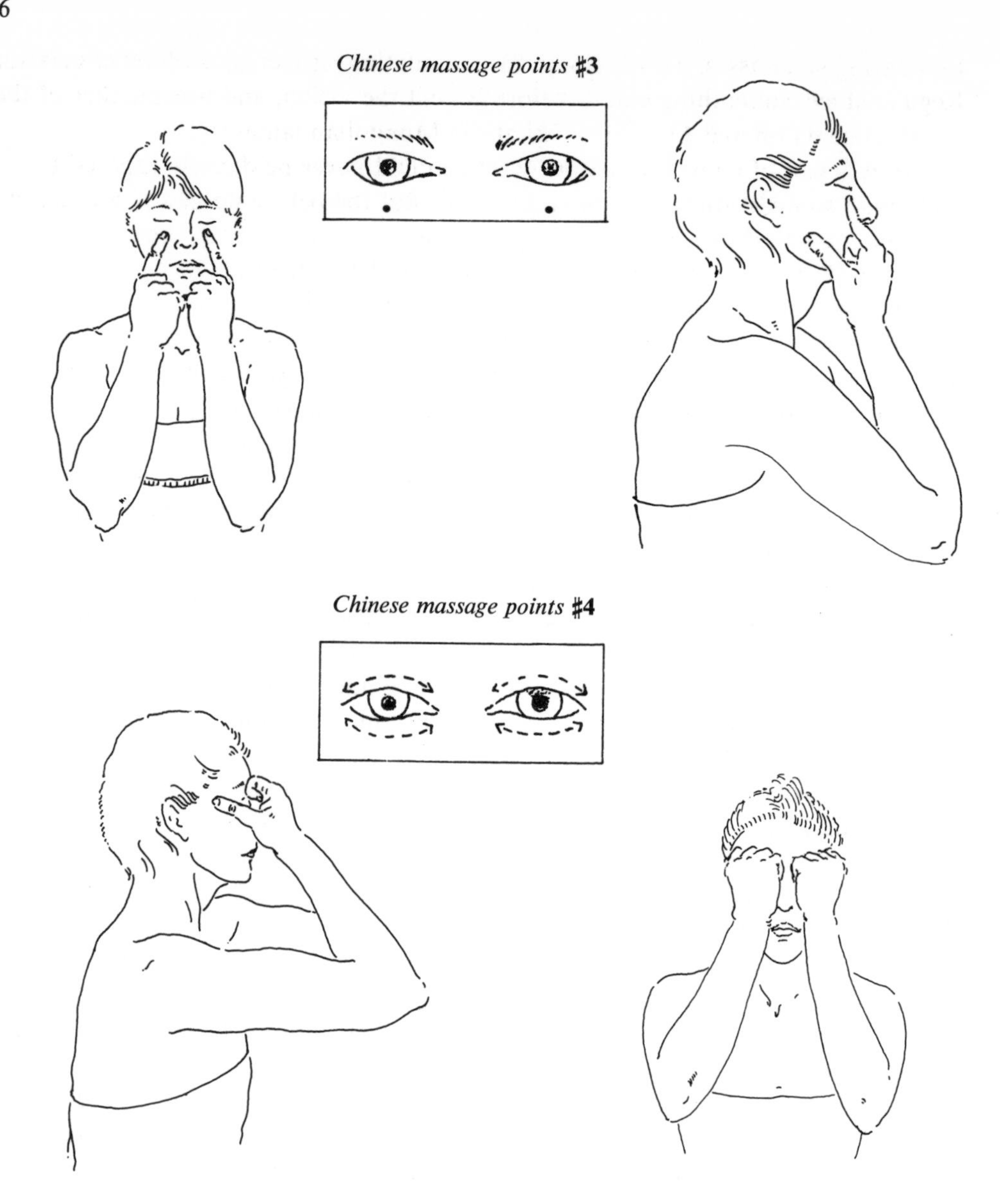

Chinese massage points #3

Chinese massage points #4

3. Place an index finger on each cheekbone, approximately an inch below the center of each eye where there is an indentation. Place the thumbs under the chin, one directly below each index finger. Simultaneously press your index fingers inward and thumbs upward to a count of sixty-four times.

4. Place a thumb on each temple and bring your fingers into fists, placing one on each eye. With the fists, rub your eyes eight consecutive times in each of the four directions—up, down, right, left.

Eye Sunbathing: Although it is commonly thought that sunlight is harmful to the eyes due to the sun's ultraviolet rays, the light of the sun, taken with discretion, is in fact wholesome to the eyes. Sunlight is not to be feared or avoided altogether by wearing sunglasses. To the body it is stimulating; it increases blood circulation. Regulated eye sunbathing can therefore benefit the vision, and was another of the eye treatments offered Masahiro Oki at the Mongolian lamasery.

The eyes can of course be sunburnt and should never be directly exposed to high-noon sunlight for longer than a second. But the before noon and afternoon sunlight can be absorbed into the eyes for longer moments. Even more so can the early morning and evening sun be taken, as can the light of the moon and stars, which provide a kind of stimulation different from that of the sun.

The following forms of eye sunbathing can be enjoyed at various hours of the day and at night. Altogether, these practices purify the eyes by increasing blood circulation to them, and are ameliorative in cases of unbalanced vision such as myopia, and eye disease such as glaucoma, conjunctivitis and sty.

1. Stand with the arms stretched outward and upward. Look into the sun (or moon or stars but not the noon-time sun) for several seconds. Gradually extend your gazing time to a few minutes.

2. Stand with the arms stretched outward and upward. Look into the sun (or moon or stars) and rotate your eyes both clockwise and counterclockwise. Keep your mind concentrated on the eye movement.

3. Stand with your eyes closed and turn your face into the sun, feeling its warmth under the eyelids. Open your eyes for a moment, the length of the moment depending upon the time of day, and then close them. Make heat with your palms and apply them gently to the eyelids. Repeat until your eyes become moist.

Water Bathing: The healthiest form of water bathing is to alternate between hot and cold water. Hot water expands and relaxes the blood vessels while cold water contracts and tonifies them. Together, hot and cold water balance the body fluids and the acid/alkaline content of the blood. To recover from the fatigue of each day, relax in a hot tub and then immerse yourself in a cold one or drench yourself under a cold shower.

To bathe the eyes only, also alternate water temperatures. Begin with a steaming cloth applied to closed eyes and follow with a cold water splashing into the eyes. Cold water, since it contracts the blood vessels within the eyes, can be used at any time to wake up the eyesight. Simply splash cold water into the eyes or using an eye cup, blink several times into cold water.

Whole-Body Purification

To abstain from food is the best means of purifying the whole body, the eyes not excepted. Whatever cleanses the blood also releases stagnation from the eyes. This you will remember was experienced by Masahiro Oki after his Sundays of fasting during childhood, but even more so did his vision profit from fasting in adult life:

> After completing my espionage training, I was assigned to an Arab country where, in accordance with the Islamic religion, the annual month of fasting was about to be observed. Islamic fasting is limited to the daytime only. At night there is feasting. But since I was accustomed to brief fasts since childhood, I chose to use the month as an opportunity to follow a prolonged complete fast, day and night.
>
> By the end of the month it was as if a cloud had been lifted from my eyes. I could see with absolute clarity. The experience taught me the values of fasting—that it aids in the purification of the blood, that it releases toxins from the body and tension from the mind, that it enables the muscles to regain elasticity. These benefits also accrue to the eyes.
>
> Nevertheless, as soon as I resumed eating, my eyesight quickly deteriorated once again. The reason, I later discovered, was that I had not exercised during the fast.

For the values of fasting to be more permanent, a fast must be practiced within the context of an active daily life that includes physical exercise. Also, the values of fasting are not only physiological. Fasting is educational, psychologically and spiritually. Basically, it is a training to control the appetite, through which the kind and amount of food best for oneself can be ascertained. Since nutrition is that element which determines how well the life-force functions, how well the individual adjusts to the vicissitudes of life, it is very useful to learn what and how much food is appropriate for oneself. Fasting can play an important role in self-knowledge on many levels. Ultimately, fasting promotes one's innate wisdom, and for this reason has been practiced since ancient times and has been, in the history of all religions, a discipline for governing the body and mind.

Nowadays, it is fashionable to go to a fasting center or spa for the purpose of becoming thinner. While there, people do lose weight and appear more beautiful. But after leaving and returning to the same life-style as before, they tend to become as before. The fasting has not been employed for self-knowledge. The state of mind has not changed. Only the body has changed and that but temporarily.

Although the first step of fasting is not eating, its purpose is not to lose weight or to change only the body. The purpose of fasting is to change the mind. The Yoga way of fasting, which is completely different from the spa method, concentrates on breaking oneself away from inappropriate habits. Its ultimate aim is to become master of oneself.

Yoga does not suggest fasting as a means of recovering disease, although it acknowledges that sickness might be alleviated through fasting. Fasting gives the internal organs a rest and frees the body of toxins which make it susceptible to disease. Also, the negative aftereffects of sickness or of an operation can be modified through fasting, in that fasting rids the bloodstream of drugs, chemicals and medications that cause imbalances in the nerve and hormone systems.

Please understand that sickness or disease or the need for an operation is a result of something; specifically, disease is the result of having provided yourself with disease-producing stimulation. This is where fasting can help tremendously: to enable you to change the stimulation you give yourself. It is true that symptoms of disease, various physiological imbalances, can be rather easily erased through a program of exercise and eating well. But if the subconscious life is not reached, that is, if psychological imbalances are not corrected, symptoms are likely to recur. Fasting can be used to unearth the subconscious, rendering it conscious. Fasting can reveal ingrained habits that they may be controlled. In this sense, fasting is a way toward mental health or becoming master of oneself. To govern oneself is to be able to change consciously the life-style, including the ways of thinking, moving, breathing, eating; to change all inappropriate stimulation to appropriate stimulation. Fasting is an excellent means to acquaint oneself with what it is to change the life-style.

Even though the Yoga way of fasting is more than theoretical (it has been used and researched for thousands of years resulting in a fund of wisdom on the subject), it is possible to make mistakes, even serious ones, when fasting. Since fasting is an extreme form of treatment, it is best practiced under the guidance of a good teacher. Any extended fast needs to be conscientiously entered upon and after the fast eating needs to be carefully resumed.

The body must be given time to adjust to a period of fasting. As preparation, you are advised to eat whole grains and vegetables only, gradually decreasing your meals from three to two to one a day over a span of two weeks. During this prefasting period, try to eliminate toxins from the body by taking an enema a few times a week and by exercising frequently. Perform breathing exercise, Yoga postures and meditation to calm your mind. Then you may begin to fast.

During the fast, continue a program of breathing exercise, Yoga postures and meditation. Make sure that your bowels move daily and if they do not, take an enema. Avoid extreme stimulations such as hot baths, saunas, or extensive periods of strong exercise. Throughout the fast, concentrate on eliminating toxins from the body and attachments from the mind.

After the fast, resume eating ever so gradually over a span of days twice as long as was the fast. Begin with very softly cooked brown rice and gradually add softly cooked in-season vegetables (roots or dried vegetables in the winter, though the onset of spring is the most likely season to fast)—as if you were preparing baby food. This is so the body may be eased and not shocked back into digestion. Continue your program of exercise, meditation and elimination, attempting to discover what is necessary and unnecessary, appropriate and inappropriate for you.

Eye Use

The worst treatment the eyes can receive is consistent one-point focusing, a use of the eyes completely averse to their naturally mobile state. You might think that in order to concentrate on something, it must be stared at. This is not so. Real mental concentration, like meditation, embodies an unstrained, effortless looking, opposite to staring. In real concentration, the muscles of the upper body are relaxed, the intake of oxygen is deep, and the exhalation of carbon dioxide is long, as when laughing.

When you need to concentrate, first relax the muscle system and put your mind at ease, detached from the subject or object of concentration. Then consciously relax your eyes and keep them moving.

Viewing Television and Film: As is widely recognized, many children in modernized countries are nearsighted. One obviously contributing factor is their habitual television viewing. Though it might seem so, the problem does not arise from the constant flickering of light emanating from the television screen. Rather, incessant staring without moving the eyes is the major cause of nearsightedness in children addicted to television. In addition, the radiation emitted by a television is detrimental to the nervous system, which of course includes the eyes.

If you are prone to television and movie viewing, you ought to know that it is less harmful to the vision to watch a comic show than a seriously dramatic one. That is because laughter, or at least a mind disposed to the kind of acceptance that levity invokes, releases bodily tension through a deepening of the breath. Deep breathing and laughter allow the viewer and his eyes to relax.

Reading: Nearsighted people sometimes come to the conclusion that their poor vision is the result of excessive reading. In actuality, the cause is not the volume of material read but the manner of reading. Myopia can result from a combination of incorrect posture and mental tension especially in the presence of poor light while reading, but it does not result from the action of reading itself.

Regarding the posture while reading, if the chin is tilted upward or pushed forward, then the back of the neck will be compressed and not allowed to stretch. Given this position, the blood cannot circulate freely up through the neck but stagnates around the shoulders and in the neck. If this neck posture becomes habitual, then the shoulder blades and spine will become misaligned, causing certain internal organs (connected to the spine by the nervous system) to malfunction. And since the power to see is related to the nervous system, to the circulatory system and to certain internal organs, particularly the liver and kidneys, the vision can be adversely affected by poor posture.

Regarding the mind while reading, if it is unpeaceful or anxious, for example if you are in a hurry to read, the eyesight can be harmed. It can also be hurt by reading simply because you cannot sleep, since insomnia is a disease involving mental tension.

Regarding the light while reading, excess intensity such as bright sunlight is even more difficult for the eyes to bear than little light. Whatever the source of illumination, and regular light bulbs are preferable to fluorescent ones, the light should not reflect directly into the eyes.

The following considerations contain guidelines to assist you in developing reading habits salutary and not injurious to your vision:

When reading a book new to you, especially one concerned with your education or profession, you might try to concentrate too hard. Such effort oftentimes results in holding the breath or breathing in spurts. A more relaxed and effective way of studying is to *read each section of new material three times over*. The first reading is a detached scanning to collect the general idea. With the second reading, your understanding will increase, allowing you to relax. During the third reading, you may partake the import of the text, now without strain.

As a study aid and also because using your hands energizes blood circulation and the functioning of the brain, *take notes while reading*. Especially take time to write down complicated or elusive words and passages.

If you are to understand deeply the content of the subject matter, it is necessary to read not with the eyes and mind alone, but with all the five senses; actually, *read with the whole body*. To do anything with the whole body is the natural way of doing. One way to insure the use of the whole body while reading is to move or sway slowly from side to side and forward and backward. This can be effected by reading aloud periodically (as Masahiro Oki learned during the Chinese language class of elementary school); for example, whenever you encounter a difficult passage or complex phrasing, *read aloud* so that the breathing will deepen and the body will sway, as happens when chanting.

Keep your eyes in motion and not glued to the text. Try looking at individual letters comprising a word and then at the full word, at big letters and then at small ones. Follow sentences individually. And to avoid blood stagnation, *periodically blink your eyes*. Also, move the book itself, taking it to distances nearer to and farther from your eyes and holding it at various heights. If you consistently keep the book at the same eye-level and distance, focusing only in one direction, you weaken the eye muscles and their power to adjust.

In general, *keep your entire situation fluent when studying*. Alternate subjects as from history to mathematics to literature, or change books dealing with the same subject.

Whatever precautions you take and whatever good study habits you develop, it is exhausting to subject yourself continuously to any one stimulation. When reading, *divert your attention* out the window or to paintings on the walls. Looking at a far point, and it is good to study in a place where you can see far into the distance, compels the eye muscles to adjust themselves. You might also have at hand a pleasing object to observe, thus refreshing your senses and nervous system. If the object is bright and shiny, it will activate the lachrymal glands, moistening your eyes.* *Interrupt your study* by listening to music or taking a walk out-of-doors. Both these activities invigorate the whole body. And at times, just close your eyes and stop thinking to give your brain a rest.

Studying can easily become a tension-producing pursuit. If it is to be a relaxing pursuit, you must aim to *keep your breath flow regular*. Try adjusting your rate of

---

* If everything around you seems too bright, the eyes are dry and possibly inflamed. To prevent dryness of the eyes, eat foods rich in minerals, sea vegetables for example, and rich in vitamins, whole grains and vegetables varying in color for example. It is recommended to obtain minerals and vitamins directly through the diet rather than to depend upon tablet supplements.

reading—speeding up and slowing down—so that your rate of breathing may find its natural, best level. The reason it is less tiring to read a popular novel than to study is that light reading automatically regulates the breathing.

Also find the posture which allows for regular breathing and good blood circulation, so as to put the least strain possible on the eyes. *Avoid stooping or hunching postures*, since these create tension in the neck and shoulders, thereby checking blood circulation to the eyes. Keep your chin pulled back and spine upright, and concentrate on putting power into the lower abdomen and lumbar region, so as to *keep the upper body free.* At frequent intervals, rotate your head and neck both clockwise and counterclockwise while exhaling; and from time to time shake your whole body suddenly all at once.

Since the neck and shoulders become particularly tight at times of overexcitement, self-consciousness and mental attachment, try to *keep your mind calm* when studying. An attitude of acceptance, better yet of enjoyment, causes the power to be centered in the lower body, freeing the upper. Also, *keep your facial expression relaxed* by letting your head nod and move, and foremost, whenever your face feels strained, just smile.

Do not worry if it takes time for your eyes to adjust to any of the various stimulations you have been advised to offer yourself when reading. Any change that causes the eyes to move is an improvement over eye fixation.

Eye Strengthening

Yoga tradition maintains at least three methods of visual concentration, which though they may be likened to staring are not harmful but strengthening to the eye muscles. Since in these kinds of "staring" the mind is to comport itself as if in meditation, there should not be eyestrain.

Visual Concentration: For the following methods, sit in a meditation posture (as illustrated on pages 134–136), keeping the spine and back of the neck stretched. In the beginning, do each exercise once or twice for a short time only. Then gradually increase your ability to focus.

1. Without blinking, focus on the tip of your nose, concentrating until tears come. Close your eyes and relax.
2. Without blinking, focus on the third eye—that point between the eyebrows—until tears come. Close your eyes and relax.
3. Pull your chin straight back to stretch the back of the neck. Without turning your head, focus both eyes on the right shoulder until tears come. Repeat with your focus on the left shoulder. Close your eyes and relax.

After practicing these exercises, cover your eyes with a cold cloth or do palm healing on them. Then perform the following eye strengthening exercises as a balance to visual concentration.

Imagination Seeing: As a preparation, lie faceup in the corpse or relaxation pose (as illustrated on pages 85–86) with the palms up and feet open to pelvis-width.

Consciously relax your whole body, breathing deeply. Then sit up and look briefly at some object. Close your eyes and continue "seeing" the object. This exercise relaxes the eyes while increasing the power of the imagination.

Meditation Seeing: Assume one of the meditation postures. Do palm healing on your eyes to relax them and then look at some object. Concentrate upon it but keep your mind empty, so as not to stare at, or attach to, the object. For variation and new stimulation, make the object a black dot drawn on white paper; or in a dark room concentrate on the light of a candle. This exercise allows the eyes to function to their maximum ability without strain, insofar as the mind is detached. It also improves one's focusing, self-hypnotic and psychic powers.

Eyesight Capacity

At one point during Masahiro Oki's employment in espionage, he was sent to live in the jungle. Since the jungle tenure followed his Zen-like training at the Mongolian lamasery plus the residency at a Yoga ashram in India, his mind and body were prepared for a wholly natural, even primitive, life-style. Of this time he said:

> In the jungle all my senses became attuned to subtle changes in the environment. No sight, no sound, no smell, no feeling escaped me. My vision became as sharp as a wild animal's. It was as if my eyes had become telescopic, able to detect details far, far away. The jungle experience revealed to me the human being's tremendous capacity to surpass supposed limitations and reach heights of sensitivity previously unknown. Nevertheless, after leaving the jungle and returning to the habits of civilized life, my eyesight collapsed once again.

The lesson, of course, is that the human being's capability varies according to his living conditions and his use of himself within those conditions. A person's level of ability, sensitivity and awareness is a direct consequence of his life-style within a particular environment.

To heal yourself, you must try faithfully and unflaggingly to better your own mental and physical condition and surrounding circumstances. This requires the courage to discipline yourself to live in a new way, a way that encourages all your senses to function to their maximum capacity. It is important to attend to the whole of you and your environment, lest your devotion be spent in vain on trying to improve one part.

Surely by now you realize that it is unhealthy to wear glasses. Take off your glasses whenever possible. Also, you surely realize that tense eyes are symptomatic of a tense mind. Learn to pacify your mind. And finally, you surely realize that your eyes are one part of the whole body, and that whatever harms the body also harms the eyes and whatever is good for the body is also good for the eyes. Create a life-style that is good for the whole body.

Chapter 3

Health and Eye Disease

Yoga teaches us to place emphasis upon the life-style; to acknowledge that habits are insinuated around the activities of daily life and that these habits affect the functioning of the whole body, its parts included. If there is some malfunction within the body, the whole life-style must be called into question to heal that part.

Nearsightedness is a malfunction within the body. It is also very likely an indication of some other, deeper illness. To improve nearsightedness you must make the whole body function normally so that this disease/symptom will disappear.

The same principles apply to healing nearsightedness as apply to healing any other chronic disease. Fundamentally, one must know the truth about the body and mind and then practice it. The truth, specifically in reference to nearsightedness, is as follows: 1) No one theory on myopia is the whole picture or entirely applicable to everyone. 2) Heredity is rarely the cause of nearsightedness. Mental and physical tension is almost always the cause of nearsightedness. Each person's source of mental and physical tension differs. 3) To heal any disease, the individual must listen to his body, becoming aware of its way of healing itself. If one's own healing power is heightened, the body will heal naturally. 4) Wearing eyeglasses makes the eyesight worse, not better, and disturbs the body's ability to heal itself. 5) Degenerative eye disease, such as glaucoma, and eyesight imbalance, such as myopia, are different phenomena with differing causes. 6) The eyes are interdependent organs related to the whole body through the nervous system.

Yoga teaches us that nearsighted people have lived in a way that has made them nearsighted. The cause of nearsightedness is to be found in the myope's daily life, and the cause of each person's nearsightedness is to some degree different from all others. To correct nearsightedness, you must reflect upon your daily life and find those practices which have contributed to your own myopia. If you did not think, breathe, move and/or eat as you do, you would not be nearsighted. The improvement of each person's nearsightedness is also to some degree different from all others. You must apply to your imbalance a way of healing that removes its cause. There is no average way of healing, though all healing necessitates changing oneself.

To change yourself, which is to change your life-style, you must regain control of your mind. To do so requires effort and training. It requires examining every part of your daily life in order to correct what needs correction.

To change one's daily life for the better, even to know what to change, it is necessary to understand the principles of natural law and to observe them. Otherwise, one may simply change from one wrong way to another wrong way or from one imbalance to another imbalance. Understanding natural law enables you to know your own mind and what you need to recover your health. Good health, once recovered, can be maintained for as long as natural law is observed.

The Properties of Natural Law

Change: Nature changes constantly. It is natural to change. The first step to health is mastering, through training, the ability to adapt to new stimulations. This entails being able to free oneself from habitual behavior. Mental attachment, which is a loss of control of one's own mind, weakens or destroys the ability to adapt. The result is a loss of balance.

To free yourself from habitual behavior, you must intentionally provide yourself with stimulations that compel you to react unconventionally, or away from your fixed patterns. Detachment training at the Oki Yoga Dojo in Japan includes corrective exercise, meditation, and fasting or a change of diet. Participation in these disciplines allows the individual to separate himself from old habits and mannerisms and thought patterns, to cut himself off from attachment.

The Japanese words for describing various Oki Yoga trainings literally convey a sense of separation from old action and undertaking new. The word for training is *gyoho. Gyo* means act or action and *ho* means the way; training is the way of action. *Danshari gyoho* is that way of action which cuts (*dan*), throws off (*sha*), and separates (*ri*) or releases from fixed behavior; the life-style of a monk is a form of *danshari. Danshin gyoho* is that way of action which cuts (*dan*) off habitual ideas or changes the mind (*shin*); meditation is a form of *danshin. Danjiki gyoho* is that way of action which cuts (*dan*) eating (*jiki*); *danjiki* is the word for fasting. Explicit in the Oriental meaning of detachment, then, is a real separation or cutting oneself off from a habitual way of thinking and acting.

Change is a powerful aspect of nature and of self-healing. Some animals change themselves by hibernation and fasting; they temporarily separate themselves from the habit of eating. Our own cells are constantly changing or want to change. Cancer cells are cells that have stopped changing; they are cells in stagnation, sick cells. Where there is a lack of change there is abnormality and then sickness.

The life-force seeks change. To become so entrenched in habits that new stimulation appears unacceptable and change is resisted is to deny the life-force. Boredom is the life-force's manner of asking for new stimulation for the mind. Sickness is the life-force's method of asking for a change of life-style. Even if your food is nutritious and you exercise, your life-style can do you harm if it persists in the repetition of habits.

Balance: Balance is the result of having the ability to adjust to change. In order to acquire balance, give yourself stimulations that are opposite to your customary ones. For example, if you are an active type of person, consciously seek intellectual stimulation; if you are a mental type of person, consciously seek physical stimulation. This is a basic principle of self-healing.

Fundamentally, healing is to turn poison into medicine or sickness into health. If a muscle is tight, it might become balanced by applying gentle stimulation. If an internal organ is weak, it might become balanced by applying strong stimulation. In general, balance calls for the application of negative stimulation; that is, the application of what is missing in the present condition. The functioning of balance, therefore, determines the nature, including the kind and degree, of any change to be made.

The aim of balance is cooperation. The highest state of physical balance arises when each part of the body is behaving normally in cooperation with all the other parts. This of course implies the unification of mind and body through a balanced nervous system. The highest state of personal or spiritual balance arises when others are cooperating with you through your cooperation with them.

Stability: Stability, which requires balance (which requires change), is that state in which the life-force functions to its maximum degree. To attain stability it is necessary to live that life suitable to you as a human being. It entails finding out for yourself the most appropriate way of moving, walking, standing and sitting; the most appropriate way of eating; the most appropriate way of sleeping. In short, you must discover a life-style fully appropriate for your needs and your capacities.

When the Chinese character for *satori* (enlightenment) is analyzed, it is seen simply to mean having one's own mind, to be in one's right mind. That is, to think in the way that is natural to you and to do in the way that is natural to you. If you habitually think and do according to the rules of other people and the rumors of society, you will tire quickly and your mind will become confused and deluded. When we imitate others, which sets up a competitiveness with them, the body and mind become unstable. Imitation is non-cooperative and leads to instability.

The highest state of stability in a physical sense is called Zen. You can reach this state through effort to develop the *hara* (lower abdomen) as the center of your being and power—the source of motivation and action. The highest state of stability in a mental or spiritual sense is called prayer or faith. You can reach this state through effort to control your mind. To have physical and mental stability is to have what is known as Buddha nature.

The way of health, therefore, according to Zen Yoga, is to be able, anytime and anyplace under any circumstances, to relax and maintain peace of mind. To the extent that you can make balance with new stimulation (transform your behavior) while maintaining stability (keeping your natural mind) is the extent to which you are healthy.

Vision and the Mind

Even so-called normal eyesight is inconsistently good. Vision depends upon not only the resilience of eye muscles or the fluidity of blood circulation. Vision is not physiological only. It also depends upon one's immediate and recurrently usual state of mind. To a tremendous extent the ability to see well is directly related to mental relaxation.

The natural tenor of the mind is to be at peace, not to be tense. But when one attaches to something, when there is strain to catch or hold onto something, the mind becomes agitated. As the mind tenses, so does the body, in particular the upper body including the neck, shoulders and arms. The eyes too are affected. When there is mental agitation, the whole upper body is tight.

Yoga teaches us to live with *mushin* (empty mind), which is a way of mental detachment that allows for receptivity, for taking things as they are. To live with detachment is to see well.

In a sense, the mind is the first organ of sight and there are at least three states of mind that prohibit normal vision, as follow:

Unfamiliarity: When the mind is confronted with something unknown, the eyesight diminishes. The opposite effect is that when the mind is familiar with its subject, because it is relaxed, the eyesight functions normally. For example, a person of science who has experience in looking through a microscope can see clearly through the lens, much better than can the uninitiated. The experienced person's knowledge, memory, and other senses assist his vision.

Indeed, when a person is seeing well it means that all his senses are keen, supporting his eyesight. This principle is frequently borne out at the Oki Yoga Dojo in Japan where many people go to improve their eyesight. As their vision improves through the dojo training, their senses of hearing, taste, smell and touch also sharpen. That is because their minds relax as do their upper bodies, their nervous systems are freed of paralysis, and becoming sensitive they once again can feel and see.

Negative Emotion: When the mind is afraid, angry, sad, or caught in some other uncomfortable emotion, it becomes tense or excited and the eyes do not see well. If one holds in mind disturbing emotions, the eyesight and ultimately the whole body suffer. This idea relates to the principle of healing that as long as the mind is negative the healing power is aborted. If a person's mentality is that of a complainer, if he persists in thinking "I am sick," he will continue to be so. Most sick people are attached to their illnesses and consider it normal to be sick. If you wear glasses, for example, you probably changelessly think of your eyesight as poor. If you never try removing your glasses, you deny yourself the possibility of ever healing your vision. But if you take off your glasses even temporarily, if you allow your mentality a positive aspect, you give yourself the chance of healing your vision.

A person whose mind is positive does not worry about his health. If he becomes sick, he does not feature the sickness but continues on with his life. When the mind is positive, one's life is active and full. But a person whose mind lingers on illness dwells in a more passive, depleted life and tends not to restore his health.

Yoga therefore teaches us that as long as we live, we live; and that as long as we live, it is natural for healing to take place. This is the most positive idea we can have about life. To live means to have the ability to heal oneself.

Mind/Body Split: When the mind and eyes are not functioning together, the eyesight is poor. For example, if you are supposedly reading a book, if your eyes are looking at the print but your mind is elsewhere abiding in some other scene or thought, you do not clearly see what you are looking at. Later if you try to recall the reading, you cannot.

The mind of a neurotic person is nearly always completely separated from his eyes. A neurotic tends to daydream and some almost ceaselessly blink their eyes. These people cannot see well. Their eyes are not in the same place as the mind.

If one sees intentionally, which is not straining to see but is seeing with concentration, the eyesight will be strengthened. For example, there are some Japanese people whose concentration is so powerful, whose mind and eyes are so conjoined, that they can write twenty, thirty, fifty, even one hundred Japanese characters on a grain of rice. Other people need a magnifying glass to discern the letters or verses or pictures drawn. A person who can write thus (and it is of course not only Japanese people who can) is seeing many points in an extremely limited space. His eyes are not staring at one point but are moving vastly over a tiny area, accordingly as his mind moves. There is unification and no paralysis.

Mind/Body Unification: To strengthen the eyesight, you must be able to maintain concentration while adjusting the focus. There are two kinds of concentration. The first is to see without the conscious mind; that is, you are looking and seeing but not intentionally noting what you see. With this kind of concentration there is no fatigue. The second kind is to see with the conscious mind; that is, you intentionally see something, perhaps observe a painting in depth wholly for the purpose of noting it. With this kind of concentration fatigue will come. It is mentally healthy and necessary to see both ways, automatically and intentionally. The Eye Training and Eye Strengthening exercises embrace both.

The principle of concentration is to try and not tense the upper body in order to see. As has been pointed out with the microscope example, that which one sees often can be seen well, because the mind is relaxed. But most people with eye trouble are already tense to some degree. As a result, when faced with a new situation, an unfamiliar place, activity or person, they become quickly tired. They cannot concentrate and cannot see because they cannot relax.

When there is mental and hence upper body tension, the memory, along with concentration and the eyesight, also fails. You surely know that if when taking an examination you are nervous, you cannot recall adequately the material no matter how diligently you have studied it. If you are tense, you miss seeing in memory even what you have seen in reality. But the memory can be revivified by closing one's eyes and relaxing.

The answer to seeing well, then, is not trying to see but relaxing in order to see.

Trying to see inhibits the vision. Making an effort to see, for example bringing an object closer to the eyes or staring at it, only weakens the eyes and aggravates myopia.

The natural condition of the human body is to change. The natural condition of the eyes is to see one point in a moment and move on, ever changing the point of focus. Unlike a machine, and the body should never be thought of as something mechanistic, the body changes of itself. When one part is disturbed or afflicted, the other parts are affected too. This is not so with a machine, whose parts are independent. If one part of a machine breaks down, the other parts do not also break down and that one part can be replaced without affecting the other parts. To stare or to wear glasses, which is to stop eye movement, is to disturb the natural condition of the whole body.

The primary ingredients of weak vision, in summary, are tension in the mind and body and a cessation of eye movement. To change weak vision into strong, you must become a master of relaxation, stop staring, and remove your glasses. If you must depend upon glasses periodically, and please realize that on and off is better than always on, then consider varying the lenses you use such as between reading glasses and distance glasses.

There are two ways of relaxing the mind and body, one active and the other passive. The active way calls for exercise and movement, the expenditure of all excess energy, an unconditional use of oneself. The passive or static way calls for rest, as in listening to music. Both ways of relaxation are needed to balance and integrate the mind and body.

The Eyes, the Mind and the Body

We have examined the relationship between states of mind and the ability to see. There is also a relationship between the mental condition and the aspect of the eyes. For example, when you are wondering or considering what to do, the eyeballs move copiously; but as soon as a decision is made, they come to comparative rest. When you are concentrating, blinking almost stops. When the mind is confused, the eyes become clouded. The eyes of neurotic people project a strange feeling; their eyes seem to be off-center and unfocused and usually the eyelids are swollen. The eyes of psychopaths are extraordinarily unbalanced; the eyes are positioned one up and one down and are of different sizes. In general, the irises of mentally ill people fade in color.

The eyes are not windows of the mind only. The whole nervous condition is revealed through the eyes. If the body becomes chilled, the pupils expand. When the body is warmed, the pupils contract. In principle what is happening is that when the parasympathetic nervous system is contracted, the pupils also contract and when the sympathetic nervous system is contracted, the pupils expand. Actually, the eyes are so sensitive to changes in the nervous system that when one feels happy, the eyelids become pink.

To diagnose your general condition, begin with observation of your eyes. A person whose nervous system and blood circulation are in good working order has clear, bright eyes with a sharp and well-defined demarcation between the iris and white of the eye. The cornea is unclouded and the expansion and contraction of the pupil runs smoothly. Blinking is also smooth. Close your eyes tightly for a while and then open them suddenly. If one eye does not open as smoothly or at the same rate as the other eye, the nervous system is unbalanced. The look of healthy eyes is healthy—shining but not overly so, which is a symptom of disease—not strained and not strange.

The eyes are very much affected by sickness. When fever is present, the eyes appear weak and the sclerotic takes on a reddish tint, producing an angry look. When there is blood stagnation, the eyes have an unclear aspect. In the case of tuberculosis, the eyeballs drop inward and the skin surrounding the eyes becomes pale. With a thyroid condition, the eyes pop outward abnormally. In apoplexy, which is primarily due to over-acidic blood and excess tension within the sympathetic nervous system, the eyeballs turn sideways toward each other. Inadequate excretion can cause a change in the shape of the eyeballs. Diabetes or nephritis (inflammation of the kidney) generally lead to cataract, and if a pregnant woman has diabetes, kidney trouble or toxemia, it is likely that the blood vessels in the eye ground will hemorrhage.

Any irregularity in the body shows up in the iris, the pupil and/or the white of the eye. The iris especially reflects one's physiological condition, and this principle is the basis of an entire branch of diagnostic medicine called iridology. Discovered in the mid-nineteenth century by a Hungarian physician, Ignatz von Peczely, and abundantly developed in the twentieth century by Peter Johannes Thiel, a German scientist, iridology holds that since the iris is comprised of numerable nerve endings, any change in the iris can be read as a barometer of the body. As shown in Diagram 5, iridology divides the iris into minute sections, each section corresponding to an internal organ, skeletal or muscular part, or function of the body. If a section of the iris changes color or becomes delineated or a spot arises thereon, sickness or imbalance is considered present in its corresponding part of the body.*

Other parts of the eye can also be read. For example, a pimple on the white of the eye or a sty on the eyelid indicates a malfunction of the liver or other digestive organ. Conjunctivitis (inflammation of the inner eyelid), keratitis, iritis, and retinitis (inflammation of the cornea, iris, and retina respectively) are all evidence of specific internal inflammation. The conjunctiva in particular, since it is wet and a suitable place for bacteria to thrive, is susceptible to infection if the blood is overly acidic. Once the blood becomes unclean, then irritation from dust, pollen or bacteria can bring on conjuctivitis. So coherent is the connection among the eyes, the mind and the body that the blood pressure in the brain can be measured through the interocular pressure.

---

* The definitive text on iridology, written by practicioner Bernard Jensen, is listed in the Suggested Reading.

Diagram 5

Sclerology chart

Right sclera

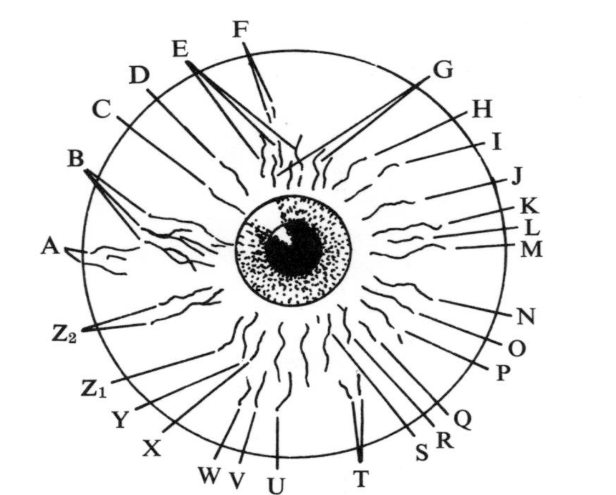

A—Bursitis or injury to shoulder (right)
B—Lung congestion
C—Neck or ear
D—Ear mastoid area
E—Sinusitis
F—Brain concussion or injury
G—Nervous disorders
H—Eye
I—Nose
J—Tonsils and throat
K—Thyroid
L—Trachea
M—Esophagus
N—Upper back
O—Middle back
P—Lower back
Q—Sciatic nerve
R—Prostate or vagina
S—Kidney and adrenal
T—Small intestine
U—Transverse colon
V—Uterus
W—Ascending colon
X—Pancreas
Y—Ovary or testes
Z_1—Liver and gallbradder
Z_2—Congestion or thorax

Left sclera

A—Thyroid
B—Upper back
C—Lower back
D—Sciatic nerve
E—Prostate or uterus
F—Hemorroids
G—Adrenal
H—Kidney
I—Small intestine
J—Colon
K—Descending colon
L—Spleen
M—Stomach
N—Heart
O—Spleen
P—History or rheumatic fever
Q—Bursitis
R—Lung congestion
S—Neck
T—Ears
U—Brain malfunction
V—Ear problems
W—Brain tumor
X—Brain concussion
Y—Sinus
Z_1—Epilepsy
Z_2—Cataracts and glaucoma

Degenerative Eye Disease

Eye diseases other than myopia, hypermetropia and presbyopia, diseases which shall be classified as degenerative, are often identified as part of the aging process. Another possibility is to consider the blood. When the blood is unclean and circulates sluggishly, the eyes are subject to corruption of vital parts.

Glaucoma: In glaucoma, the interocular pressure is excessively high, resulting in pathological change. The eyeball becomes hard and the pupil greenish in color. Among the early symptoms are bloodshot eyes, blurred or steamy vision, and loss of peripheral vision. As glaucoma progresses, there can be damage to the retina or even destruction of fibers of the optic nerve, resulting in blindness. Apparently because of unsuccessful treatment or lack of treatment, more than half of the cases of adult blindness are due to glaucoma.

Modern medicine, which considers the cause of glaucoma unknown, treats it by attempting to lower the interocular pressure. Drugs given as eye drops to contract the pupil are prescribed for daily ministration. When drugs fail, surgical procedure is undertaken, usually boring a hole in the sclerotic. Eyeglasses, admittedly proven ineffective for glaucoma, need to be adjusted frequently in lens power or are simply not prescribed.

According to the viewpoint of Yoga, the underlying cause of glaucoma is poor circulation of poor quality blood within the entire body. High blood pressure is never limited to the eyes only. Blood congestion within the eyes forces a breakdown of their purification mechanism. This mechanism functions by means of the vitreous fluid encased in the eyeball. When flowing normally, the fluid supplies oxygen and nutrients to the surrounding parts of the eye and accepts waste. But if more fluid enters than leaves the center of the eyeball due to blood congestion, that is, if the acceptable level of fluid within the eyeball is exceeded, the interocular pressure rises.

To improve glaucoma there must be purification of the blood through a complete change of diet and improved exhalation. The supply of oxygen to the circulatory system and especially to the eyes must be increased. For this, whole-body and eye exercises are recommended.

Cataract: As with glaucoma, cataract is most prevalent among elderly people. It is a disease in which the crystalline lens (usually of one eye at a time) becomes foggy and dull white in color. Light entering the pupil is then diffused by the muddiness of the lens, impairing the vision. Early symptoms of cataract are double vision and the eyes tiring easily when reading. As cataract advances, the eyes cannot tolerate light. Objects under bright light appear too bright, and shiny objects appear too shiny. Yet it is characteristic of cataract for the vision to vary from time to time.

According to modern medicine, the cause of cataract is uncertain though it is associated with diabetes, with various occupational hazards such as overexposure to heat or radiation, and with eye injury. In general, it is considered a result of growing old and in the majority of cases is called "senile cataract." There are no drugs to treat it. It is handled surgically by removing the lens. Afterward, thick-lens glasses are prescribed. However, since the life-style has not been altered, cataract usually develops in the other eye in due time.

Yoga traces cataract to poor blood quality. In senile cataract, the blood is not congested and the condition is generally painless. Undoubtedly, however, there is a history of chronic disease such as diabetes, nephritis and/or constipation behind cataract. The body with cataract is one that is not eliminating waste and not renewing itself adequately. This abnormality accrues to the crystalline lens.

The crystalline lens is composed of fiber which can deteriorate due either to excess liquid within the lens or to old fiber adhering to the lens. Even though the lens fiber renews itself annually, undischarged deposits can stick there, hardening and becoming yellow. Since the lens is that part of the eye which gathers light from the pupil in order to focus a picture on the retina, if the lens becomes muddy with excess fiber or water, the picture becomes cloudy. At this point, cataract (opacity of the lens) forms.

To improve cataract, the blood must be purified by balancing its acid/alkaline content, the kidneys healed, and the processes of elimination normalized. In other words, the sources of cataract can be traced to the life-style, including diet deficiency, misuse of the body and negative mental attitude.

Detached Retina: When small pieces of the retina become separated from the choroid, the middle lining of the eyeball, there is detached retina. Its catalyst is a change in shape of the inner eyeball, the vitreous body, which cushions the retina. If the vitreous body shrinks (due to myopic conditions, for example), then fibers comprising the retina are pulled from the neighboring lining, the choroid, which is the eyeball's blood supplier. In detached retina, the vision first becomes dull, spots are seen in front of the eyes, and blindness can result. Although a somewhat rare disease, when detached retina is treated by modern medicine the procedure is a delicate surgery involving cauterization to produce scar tissue. The aim is for the scar tissue to pull the retina back into its normal field.

Amaurosis: In amaurosis there is loss of sight without apparent lesion of any part of the eye, although it usually includes retinitis. Since pathology is unapparent, medical science attributes amaurosis to some defect of the optic nerve, the brain or the spine. From Yoga's viewpoint, it might also be attributable to an imbalance of those internal organs related to the functioning of the eyes.

It must be obvious by now that Yoga does not necessarily subscribe to customary theories on sight imbalance and degenerative eye disease. Yet Yoga neither condemns nor dismisses modern medicine. Rather, it utilizes what is appropriate. Especially noteworthy from the Yoga viewpoint are modern medicine's abilities to conduct eye examinations and diagnose disease scientifically.

Nevertheless, the purpose of Yoga is not to recover disease. Realistically, the eyes should be taken care of before symptoms of decline appear. But if symptoms such as weakened vision, tired eyes and headache do appear, please think of them as warnings from the life-force that the life-style is hurting the eyes. It is at this point, the very tenuous yet potentially debilitating first stage of disease (since disease tends to worsen progressively), that Yoga suggests you step in firmly and commence correcting the life-style immediately.

Improving Eye Disease

The inability to improve disease is directly related to an inability to discern its cause. In other words, ignorance is the basis of incurability. Insofar as the source of an eye disease is known, its recovery can be known. Sight imbalance, it has been stated, is fundamentally caused by habitual mental and physical tension, so that where there is myopia there is a paralyzed nervous system. It has also been stated that the fundamental cause of degenerative eye disease such as glaucoma and cataract is impure, stagnated blood.

For treating nearsightedness, you must find the source of your particular tension. In the case of glaucoma or cataract, the source of the imbalanced blood must be found. Then you are advised to strike your own path. Just as one kind of food or herb will somehow affect you differently than it affects others, any one method of vision improvement will help you more or less than it helps others. Your path needs to be a way of life that is natural for you and different from the unnatural life-style that caused the mental and physical tension or the stagnant blood in the first place.

As an example of a self-conducted program for vision improvement, please hear the case of a university student who went to study at the Oki Yoga Dojo in Japan. His vision was calculated to be approximately 20/400. From Masahiro Oki he received individual corrective exercise for nearsightedness, which he practiced religiously each day. He assertively participated in the dojo training,* ate only the food provided, and within a week his eyesight had improved to approximately 20/30.

His case might lead you to wonder if you also, by practicing his corrective

* The training at the Oki Yoga dojo in Mishima, Japan includes daily classes in breathing, purification and strengthening exercises, running, Yoga postures, group chanting and meditation, hot/cold bathing, participation in care of the dojo, and sessions for lecture and discussion and questions and answers. Other frequent activities are hiking, classes in martial arts (especially aikido), shiatsu massage, Japanese cultural arts, and food preparation. Limited free time is available for study, for reflection, for journal writing, for practicing individual corrective exercise, or to receive treatments of acupuncture and moxabustion. All participants are served the same kind and amount of food, except for those in a period of pre- or post-fasting for whom special food is prepared; fasting is supervised. Corrective exercise, if requested, is designed for the condition of the individual and is also supervised. There are frequent day-trips to a hot-spring dojo and monthly parties.

exercises and disciplining yourself along the lines of the dojo training, could cure your vision to the same extent that he did in one week. The answer is no. What were his imbalances are not necessarily yours.

The university student's problem was tension due specifically to incorrect posture and breathing. The corrective exercises designed for him addressed his body, his posture, his breathing difficulties. Also, before arriving at the dojo, he had already improved his diet, first by fasting and then by choosing to eat natural foods in balanced proportions. Additionally, he had begun a practice of mental relaxation to free his mind of attachment to old habits. Because of these prior endeavors, his body was primed to accept and utilize totally the corrective exercises tailored to his condition.

What can be learned from the case of the university student is the value of applying to one's daily life the theory of corrective exercise; that is, to discover one's imbalances and change them consciously. The aim of corrective exercise is to create a stable balance through intentionally releasing tension where it is felt and adjusting the breathing when it is inadequate. Corrective exercise (the subject matter of Chapter 5) was formulated by Masahiro Oki to enable the life-style of the individual to reflect the properties of natural law. Corrective exercise is thus a vital component of the healing process.

Chapter 4

Holistic Vision

Myopia is founded in an unnatural life-style, including an abnormal use of the eyes themselves. Most people are ignorant of how to use their eyes, indeed their body and mind, correctly. Most people do not know the life-style appropriate for them and never find out. Most nearsighted people die myopic without ever having been aware of the cause of their nearsightedness.

It is important to improve the functioning of the whole body and mind, not only of the eyes, to correct nearsightedness. That is because the functioning of the eyes is closely related to the functioning of other organs of the body, especially the kidneys and liver. Also, the use of the arms, legs and neck is connected to the use of the eyes. The balance of the vision is traceable to the balance of the spine and the diet. You have already seen how the eyesight can be affected by mental and emotional negativity. In short, any excess mental or physical tension can disturb the vision.

No single element of one's life is easily identifiable as the initiator of myopia. Many, many factors, physical, mental and spiritual, affect the balance of the whole body and hence the eyes. You must discover the primary cause or causes of the mental and physical tension leading to your nearsightedness and then experiment with ways to free that tension. To arrive at the source of your nearsightedness, please study those factors known by Yoga to affect the vision.

Eyesight and the Internal Organs

Most people suffering from nearsightedness, which, please remember, is a chronic disease, also suffer, probably unknown to them, from some disease or imbalance of certain internal organs. For example, at the Oki Yoga Dojo in Japan, when a person afflicted with gastroptosis (dropped stomach whereby the contractive power of the abdomen is diminished) or with Bright's disease (kidney inflammation) rids himself of the internal ailment, his vision improves. The reason that the eyesight improves through the practice of Yoga is that Yoga normalizes the blood circulation and heals the internal organs, some of which are absolutely connected to the health of the eyes.

To make the eyes strong, you must make the internal organs strong. The same principle applies to the heart. Although the heart pulsates with an abundance of blood, it takes its own nutrition from other organs and for strength is dependent upon the well-being of the liver and lungs. For nutrition, oxygen and purification, the eyes are also almost completely dependent upon other parts of the body.

It is possible, however, that you are unaware of the actual condition of your internal organs, and you cannot always depend upon a physician to describe accurately your condition. Nevertheless, you can diagnose certain imbalances by yourself.

When you realize that the autonomic nervous system (which is intimately associated with the central nervous system) extends from the vertebrae to the internal organs, it becomes apparent that the functioning of internal organs is contingent upon the entire nervous system. And of course the optic nerve travels to the eye from the central nervous system.

The autonomic nervous system innervates all smooth muscle tissue in the body, the heart and the glands, and controls the body's most elementary functions. Since it responds to the stress of feelings and emotions, one part is designated as the sympathetic nervous system and another part, generally complementary/antagonist, the parasympathetic nervous system.

Both sympathetic and parasympathetic fibers lead to most organs. How the organs react depends on which part of the autonomic nervous system is dominant at the time. For example, when the sympathetic impulses are in the ascent, the heart beats faster, the pupils of the eyes dilate, the blood pressure rises, the contractions of the gastrointestinal tract are decreased. The drug atropine (sometimes administered to myopes as previously stated) to a certain extent paralyzes the parasympathetic nervous system, leaving the sympathetic nervous system in control. When this drug is taken, therefore, the heart beats faster, the pupils dilate (allowing for temporary far sight), the blood pressure rises, and so forth. When the parasympathetic system is dominant, the opposite phenomena occur: the heart beat slows down, the pupils contract, and so on.

Most of the innumerable reflex acts by which the body functions are controlled by the autonomic nervous system; that is, at the level of the spinal cord. The ganglia, or nerve centers, of the autonomic nervous system are strung along the spinal column extrinsic to its bony structure. Thus if the spine is unbalanced, if any of the vertebrae are out of line (too far up, down, to the right, to the left, in or out), then the organ or part of the body connected by the atuonomic nervous system to that vertebra is under stress and somehow unbalanced. The imbalance can arise in the spine, due for example to the posture or an injury, or in the organ, due for example to the diet. In either case, the imbalance will soon enough be mutual, manifesting in spinal sensitivity and disturbance in the corresponding organ or part of the body. And when one organ is unbalanced, its most related organs, indeed all the body, are affected through the entire nervous system.

For the purpose of self-diagnosis, the following chart lists the connections between each vertebra and its related organs, parts, and/or functions. Please refer to the spinal column as illustrated in Diagram 6 to locate each vertebra under consideration.

Diagram 6

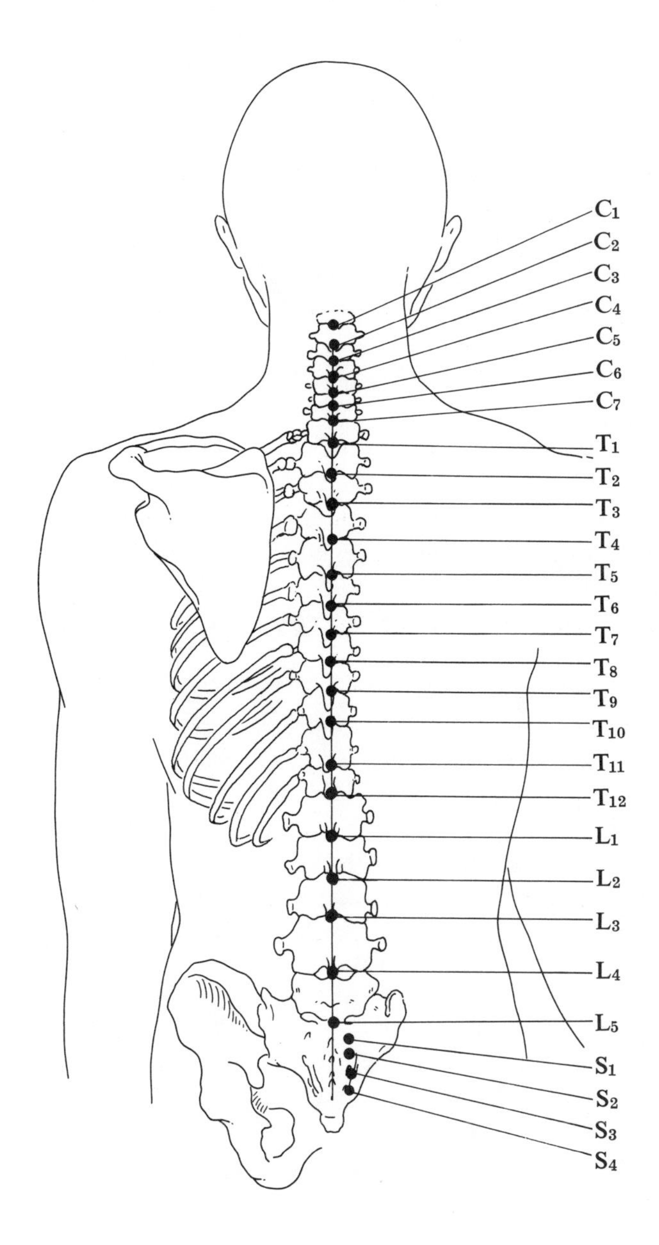

Cervical (neck) Vertebrae

C_1—eyes and ears

C_2 and C_3—blood circulation to the brain

C_4 and C_5—eyes, ears, nose and diaphragm

C_5 and C_6—thyroid gland, throat and heart

C_6—throat and arms

C_7—heart and vagus nerve (parasympathetic functions)

Thoracic (back) Vertebrae

T_1—trachea, heart and arms

T_2—trachea

T_3—lungs and heart

T_4—liver and lungs

T_5—lungs, eyes and pylorus (opening from stomach into duodenum)

T_6—diaphragm, pleura (membrane enveloping lungs) and stomach

T_7—diaphragm, kidneys and contraction of adrenals

T_8—diaphragm, pancreas and liver

T_9—spleen, pancreas and bladder

T_{10}—kidneys and small intestine

T_{11}—stomach, intestines and uterus

T_{12}—large intestine and uterus

Lumbar (small of the back) Vertebrae

L_1—bladder, and contraction of stomach, intestines and liver

L_2—contraction of stomach and intestines, and appendix

L_3—sexual organs and legs

L_4—bowel movement, large intestine, legs, uterus and bladder

L_5—bladder and legs

Sacral (lower end of the hips) Vertebrae

S_1—bladder, sexual organs and legs

S_2—sexual organs

S_3—anus and bladder

S_4—bladder, anus and legs

S_3—anus and bladder

S_4—bladder, anus and legs

Through the application of finger pressure, as in shiatsu massage, to a particular vertebra, the condition of its related organs, parts and/or functions can be determined. Discomfort indicates trouble. For example, if you experience pain, hardness or tightness when the area surrounding the fourth thoracic vertebra is stimulated, then the liver, and perhaps the lungs too, are unbalanced. Corrective exercise, which adjusts the breathing and posture, and a change of diet are natural means to align the spine and heal unbalanced parts of the body.

The Eyes and the Kidneys: The kidneys are involved in filtration of the bloodstream to remove impurities. When they cease to function properly the blood pressure rises, pointing toward cardiovascular degeneration and hardening of the arteries. The interocular pressure rises too, potentially leading to glaucoma.

Nearsightedness in particular can be associated with weak kidneys, which is why Yoga recommends that a myopic person purify his blood through fasting, diet adjustments, exercise and deep breathing in the case of myopia. When the blood is clean, the kidneys are not overworked and the supply of oxygen and nutrients to the eyes is increased.

Since the kidneys also help regulate hormone secretion and thus control the body's vital energy, they are extremely sensitive to stress. Their aim is to monitor it, but if a person sustains prolonged stress the kidneys undoubtedly weaken. This depicts another link between mental tension and nearsightedness. According to Oriental intuitive thinking, the kidneys store vitality inherited from one's ancestors; therefore, if a child manifests poor eyesight, it is generally concluded that his mother suffered ill health during pregnancy.

This relationship between the kidneys and the eyes by virtue of the fluid (blood and hormone) systems of the body is stressed in the study of traditional Oriental physiognomy. The nervous system also figures in connecting the kidneys with the eyes, since the kidneys influence the power of the pupil to expand and contract. Thus, in physiognomy, kidney imbalance is diagnosed through such facial indications as dark circles or purple bags appearing under the eyes.

The Eyes and the Liver: The liver works in cohesion with the kidneys to detoxify the blood and maintain a substantial level of energy for undertaking action. All harmful substances in the bloodstream, on the way to the heart from the intestines and stomach, are supposed to be abstracted by the liver. But if the liver becomes overburdened, especially due to toxic diet or overeating, the liver will weaken and the blood not be kept clean. Blood circulation will be inhibited, and since circulation in the liver parallels circulation in the eyes, weak eyesight or degenerative eye disease can emerge.

The liver also acts as a storehouse for enzymes, vitamins, minerals and hormones. If the liver becomes unbalanced, the store of vitamins A and D, both essential to clear vision, will be diminished.

In Oriental physiognomy, liver problems are detected in the whites of the eyes becoming cloudy, yellowish or reddish. People who drink alcohol excessively have yellowish colored eyes, as are the eye-grounds of persons frequently worried or angry. A "liver type" person, in Oriental diagnosis, is characterized as stubborn, hard working, overly indulgent, and in possession of seemingly endless drive that readily leads to fatigue. Liver imbalance exhibits itself physiologically as stiffness in the joints and muscles. Given the combined mental and physical tension of a liverish person, his eyes too are prone to rigidity.

There are several ways to care for the liver, all of which relate to the life-style, as follow: 1) Foods which tax the liver, prohibiting adequate blood absorption and circulation, should be avoided; such foods are animal fats, oils, salted or high sodium content foods, and alcohol. Artificial preservatives and medical or recreational drugs also harm the liver. 2) Overeating damages the liver as does constipation. It is recommended to eat moderately and practice fasting from time to time. To assure waste elimination, eat whole grains, especially brown rice, buckwheat noodles, fresh green and root vegetables, and sea vegetables. 3) Emotional instability brings on poor blood circulation, affecting the liver. To balance oneself emotionally, it is recommended to practice deep breathing, methods of physical and mental relaxation including meditation, and to offer oneself stimulations that invoke joy and pleasure.

The Eyes and the Stomach: The stomach is directly influenced by emotion. The color of its lining changes in the same way that one's face color changes according to the emotional state. Similarly, the eyes and stomach can be seen to work cohesively. When a person is hungry, the expression in his eyes becomes glaring or glittery; when he is overly full, the eyes become dull or glazed. An unbalanced appetite or chaotic eating is reflected in the eyes to the extent that a person with such a habit is prone to myopia.

If the liver or kidneys, and more especially the stomach, weakens, the posture begins to stoop. The position of the shoulder blades becomes unbalanced, and since those nerves controlling the ciliary muscles are located around the first thoracic vertebra, the vision suffers. When the shoulder blades are out of line, the shoulders and neck become tight, paralyzing the nervous system and decreasing blood circulation to the eyes.

The Eyes and the Intestines and Reproductive Organs: Constipation as well as sickness in the reproductive organs will undermine the blood quality in the lower abdomen. When the lower abdomen or *hara* power weakens, the upper body (often the misplaced seat of power or contraction) becomes congested, affecting the vision. After a bowel movement, you might have noticed, the eyes feel clear, indeed the whole head does.

If the lower abdomen tenses or is upset due to sexual frustration, the eyes too will tighten and the upper nose becomes congested. Eye problems of teenagers can frequently be traced to sexual tension or constipation. In children, the cause of sight imbalance is usually digestive problems, generally due to incorrect diet on the part of the child's mother during pregnancy.

The Eyes and the Brain: If the mind is overly excited, or if one is in an intense mental state, the eyes reflect the brain's stress, exhibiting blood congestion and ready fatigue. Since it is so essential to one's eyesight to be able to balance and quiet the mind, recommendations for increasing the flow of blood and oxygen to the brain are offered as follow:

1. Practice the Yoga Shoulder Stand Pose (as illustrated on page 78) and sometimes wear a headband round your forehead.
2. If it is necessary to do prolonged study or desk work, through which the power of the lower abdomen becomes sluggish and the upper body tight, then periodically practice deep breathing to free the abdomen and expand the lungs. Even while sitting, remember that the lower abdomen is your center of gravity; try and keep your power there, not letting it slacken or shift to the shoulders and neck. Interrupt your sitting time by taking a walk to relieve blood congestion and tired eyes.
3. If fatigued from standing extensively, or if some parts of the body have been held in check due to a work posture, or if you feel light-headed, then allow your body to move about freely. Let the body say how it wants to move in order to free accumulated tension. Simultaneously, yawn, stretch and inhale deeply, taking plenty of oxygen to relax your muscles and clear the brain.
4. Whenever possible, consciously utilize your hands in efficient ways. The functioning of the hands is related to the functioning of the brain, and the eyes are involved in the coordination of both.

Yoga Poses Plus Vision Stimulation

By understanding the relation between the eyes and internal organs, you can acknowledge that your vision is not an independent matter but depends upon the condition of the whole body and mind. To improve the functioning of the eyes, it is important to train yourself in a way that integrates the body and mind. You can first reflect upon your posture, breathing, eating habits, life-style to discern what is inappropriate and unnecessary for you. You can then strive to keep your spirit light and joyous, to meditate and to exercise.

You might wish to begin a program of exercise with the practice of Yoga *asanas* (poses) for healing the internal organs. When these poses are combined with specific eye movements as outlined below, the vision will benefit doubly. Please realize though that whatever internal organ dysfunction exists will be attended to by the body before the eyesight, since the eyes are secondary organs of survival.

Asanas: The following Yoga poses may be performed at any time of day, but should not be practiced until one or two hours after eating. Do some warming-up exercise beforehand.

1. Triangle Pose

Stand with the legs open twice shoulder-width and the feet turned toward the right. Look straight ahead.

Inhale, raise the arms to shoulder level and open the chest by bringing the shoulder blades together behind you. Exhaling, stretch both knees and raise the left arm straight up while bending the torso to the right, letting the right arm slide down to the right foot. Look up at the left hand. Hold the posture through a few complete breaths.

Inhale and return to center. Turning the feet toward the left, perform the movement on the opposite side.

Repeat the entire sequence a few times, twice as many times on the more difficult side.

Effects: Improves the flexibility of the chest, abdominal and side muscles. Stretches the spine and improves the posture. Increases peristalsis of the intestines.

2. Triangle Twist Pose

Stand with the legs open twice shoulder-width and the feet turned toward the right. Look straight ahead.

Inhale, raise the arms to shoulder level and open the chest. Exhaling, twist the upper body to the right and touch the left hand to the floor in front of the right foot. Look up at the right hand. Hold the posture through a few complete breaths.

Inhale and return to center. Turning the feet toward the left, perform the movement on the opposite side.

Repeat the entire sequence a few times, twice as many times on the more difficult side.

Effects: Stretches and balances the legs, stimulates the lumbar area, and opens the chest.

3. Lateral Angle Pose

Stand with the legs open twice shoulder-width and the right foot turned outward. Look straight ahead.

Inhaling, bend the right knee into a right angle and place the right hand behind the right foot. Keeping the left leg stretched, place the left hand on the left thigh. Exhaling, stretch the left arm along the head, in a straight line with the stretch of the left leg. Hold the posture through a few complete breaths.

Inhale and return to center. Turning the left foot out, perform the movement on the opposite side.

Repeat the entire sequence a few times, twice as many times on the more difficult side.

Effects: Stimulates blood circulation to the ankles and knees. Strengthens the digestive organs, and opens the chest.

4. Cobra Pose

Lie facedown with the legs together, toes pointed, and forehead on the floor. Bend the elbows and place the palms facedown near the upper chest. Bring the shoulder blades together behind you.

Inhaling, slowly lift the head and upper chest by means of your spine and the inhalation, not by means of your arm power. Continuing the inhalation, further lift the upper body without the use of arm power. Inhaling fully, push the palms and pelvis into the floor, and, now using your arm power, lift the torso all the way up, arching backward. Hold the posture, with the chin elevated and the chest fully opened, through a few complete breaths.

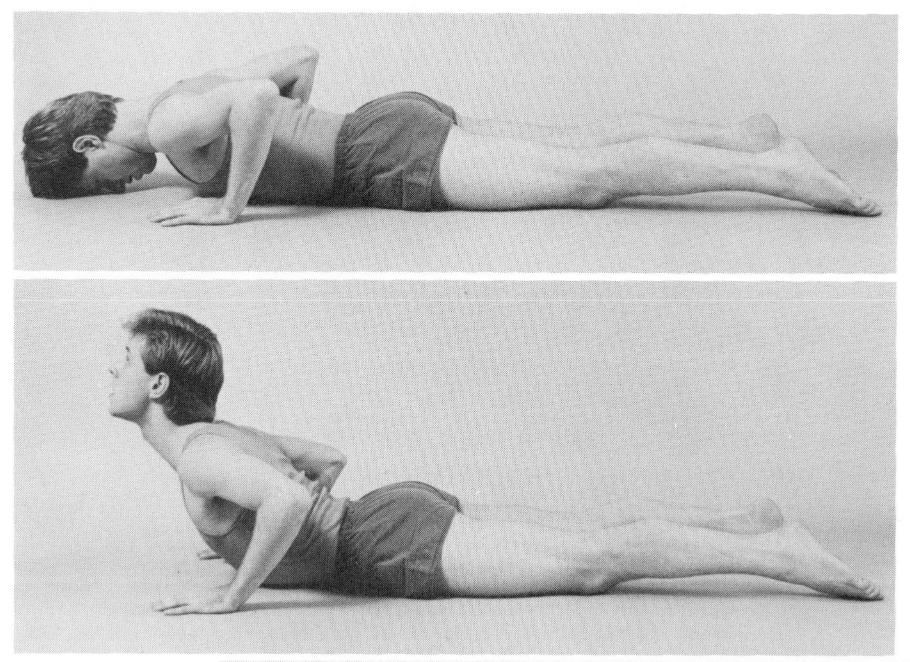

Inhale. Exhaling, turn the head slowly to the right, letting your eyes follow the head movement. Inhale. Exhaling, perform the movement on the opposite side. Inhale and return to center.

Exhaling, come down slowly and relax.

Effects: Strengthens the spine and helps maintain an upright posture. Relieves backache, especially pain related to the intestines. Stimulates the pelvic area, including the reproductive organs.

5. *Locust Pose*

Lie facedown with the legs together, toes pointed, and chin on the floor. Make fists and place them under you at the pubic bone. Keeping your forehead on the floor, stretch the back of the neck.

Inhale deeply and hold the breath in the lower abdomen for a few seconds. Exhaling, push the fists into the floor while raising the pelvis and legs. Hold the posture for as long as possible while breathing.

Exhaling, come down slowly and relax.

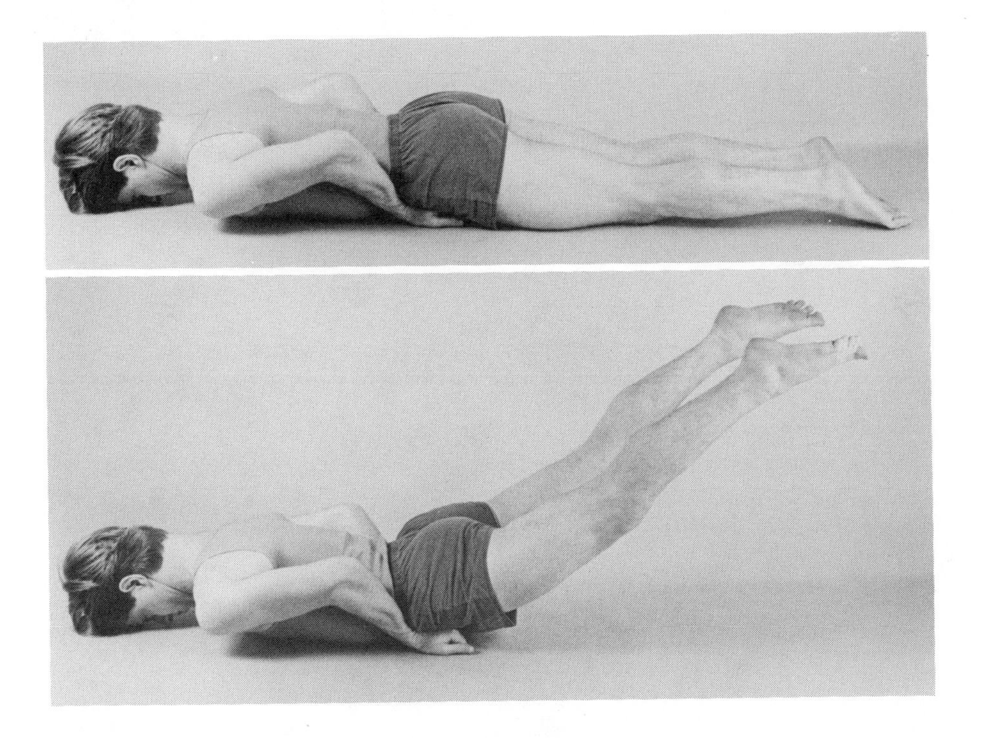

Effects: Strengthens the lower back, pelvis and abdomen. Stretches the upper back, shoulder and arm muscles. Increases circulation to the chest area, benefiting the heart, lungs and liver.

6. Shoulder Stand Pose

Lie faceup with the legs together, Achilles' tendons stretched, and arms stretched at your sides, palms down. Pull the chin in. Throughout the upward sequence, follow the movement of the legs with your eyes.

Inhale. Exhaling, raise the legs to a ninety degree angle to the body. Inhale. Exhaling, bring the legs overhead, supporting the lumbar region with your hands,

and rest on the tops of your shoulders. Inhale and bend the knees. Exhaling, straighten the legs all the way up, and push the chin into the chest.

Inhale. Exhaling, stretch the Achilles' tendons while rolling your eyes up and down and then side to side. Inhale. Exhaling, point the toes and relax your eyes. Hold the posture for a minute or two while breathing.

Exhaling, come down slowly, going backward through the sequence of movements, and relax.

Effects: Strongly stimulates the thyroid and parathyroid glands, thereby helping to balance the organs of digestion, elimination and respiration. Increases blood circulation to the liver and kidneys, and increases vitality.

7. Fish Pose

Lie faceup with the legs together and Achilles' tendons stretched. Bend the elbows.

Inhale. Exhaling, open the chest, arch the spine up, and come onto the crown of your head. With your weight on the head and elbows, hold the posture through a few complete breaths.

Inhale. Exhaling, push the chest further up, turn the knees inward to touch each other, and stretch the Achilles' tendons. Roll your eyes up and down several times, breathing naturally. Roll your eyes side to side several times, breathing naturally.

Exhaling, come down slowly and relax.

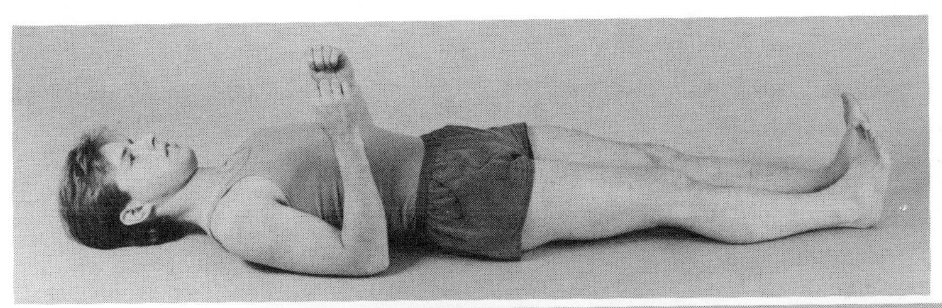

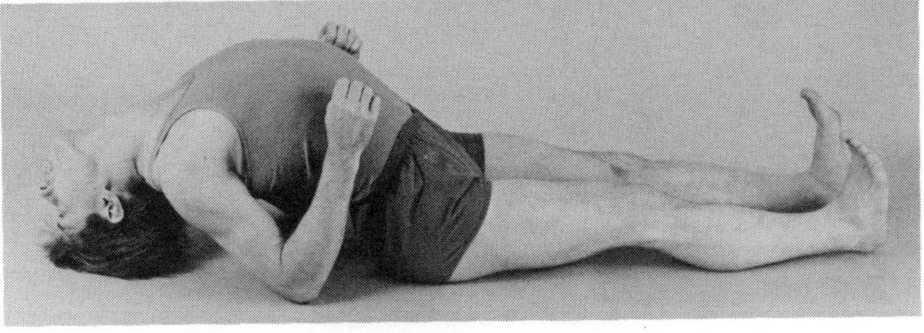

Effects: Expands the chest and stimulates the lungs. Massages the neck and shoulders, and strengthens the waist and spine. Increases blood circulation to the cervical and lumbar regions.

8. Plow Pose

Lie faceup with the legs together, Achilles' tendons stretched, and arms stretched at your sides, palms down. Pull the chin in. Throughout the exercise, follow the movement of the legs with your eyes.

Inhale. Exhaling, raise the legs to a forty-five degree angle. Inhale. Exhaling, raise the legs to a ninety degree angle, perpendicular to the floor. Inhale. Exhaling, lower the toes to the floor, keeping the Achilles' tendons stretched. Hold the posture through a few complete breaths, bending the knees with the inhalation and stretching them with the exhalation.

Exhaling, come down slowly and relax.

Effects: Stretches the spine and all the back muscles. By compressing the abdomen, increases blood circulation to the abdomen and legs. Helps balance the liver functions.

9. Bow Pose

Lie facedown with the forehead on the floor.

Inhaling, raise both legs and catch the ankles. Exhaling, lift the upper body while pulling the feet toward the head. Holding the posture, with the neck arched backward and the shoulder blades together behind you, look up.

Inhaling and exhaling, rock the body side to side, down on the exhalation and up on the inhalation, through several complete breaths. Throughout the rocking, follow the movement of the upper body with your eyes.

Exhaling, come down slowly and relax.

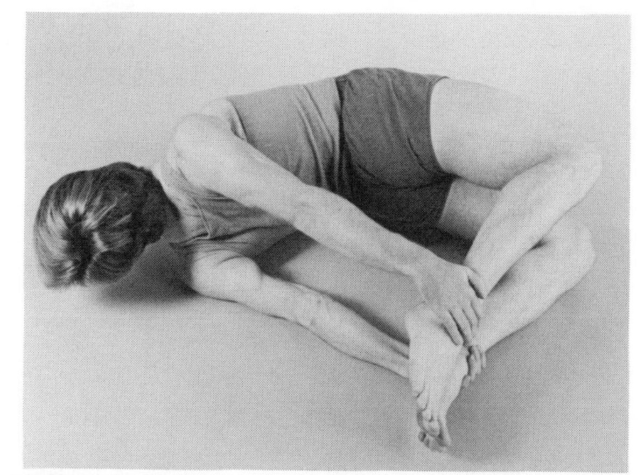

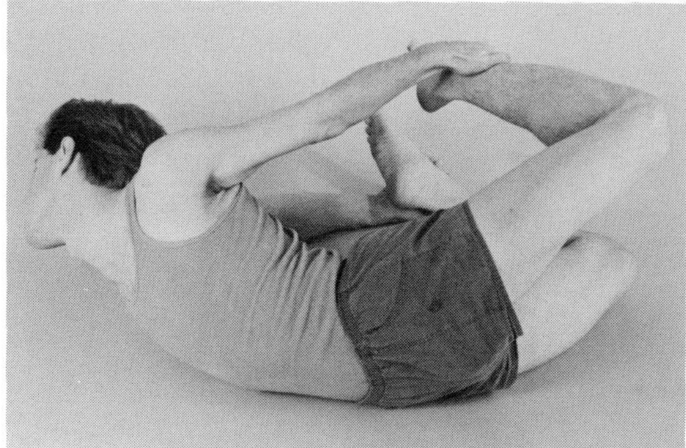

Effects: Reaps the combined effects of the cobra and locust poses. Stretches the abdominal muscles, massages the internal organs, and stimulates all the vertebrae.

10. Wheel Pose

Lie faceup with the knees bent and the feet open pelvis-width. Place your hands palms down on either side of your head, the fingers pointing toward the feet.

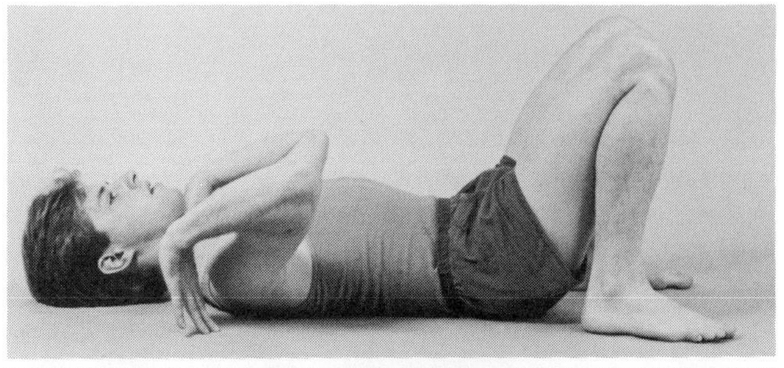

Inhale. Exhaling, push the pelvis, chest and shoulders straight up, rolling onto the top of your head. Inhale. Exhaling, push the torso to its maximum elevation.

Inhale. Exhaling, roll your head slowly from side to side, following the head movement with your eyes. Hold the posture, continuing the eye exercise, for as long as possible while breathing naturally.

Exhaling, come down slowly and relax.

Effects: Stretches the muscles of the legs, hips, shoulders and arms. Stimulates all the vertebrae. Relieves slumping postures, and improves the functioning of the heart and liver.

11. Forward Bend Pose

Sit with the legs together and the hands together, both outstretched in front. Pull the chin in and look at your thumbs.

Inhale and lean backward to a forty-five degree angle, keeping your eyes on the thumbs. Exhaling, bend forward slowly from the lumbar region, keeping the arms outstretched. Stretch the Achilles' tendons and catch your toes to pull yourself further forward, trying to touch the chest to the knees. Close your eyes and hold the posture through a few complete breaths.

Inhaling, come up slowly and relax.

Effects: Stretches the neck, waist, leg, knee and feet muscles, and the spine. Tones the kidneys, liver and pancreas. Increases blood circulation to the intestines, bladder and reproductive organs.

12. Cat Pose

Kneel on all fours with the knees pelvis-width and the hands in a line with the shoulders, fingers pointing forward.

Inhale. Exhaling, push the lumbar region down while elevating the chin. Look toward the ceiling and hold. Slowly exhale and return to center. Inhale. Exhaling, push the lumbar region up while pulling the chin into the chest. Hold the posture. Slowly exhale and return to center.

Repeat the entire exercise several times.

Effects: Affords spinal flexibility and balances the shoulder blades. Improves blood circulation to the abdomen.

13. Hara Strengthening Pose

Stand with the legs open pelvis-width and the feet parallel. Pull the chin in. Open the chest by bringing the shoulder blades together behind you and relax the shoulders. Make a loose fist of each hand, placing the thumb over the fingers. Push the

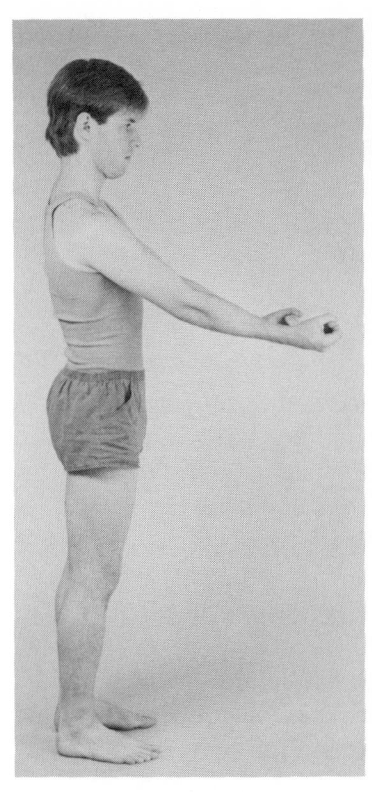
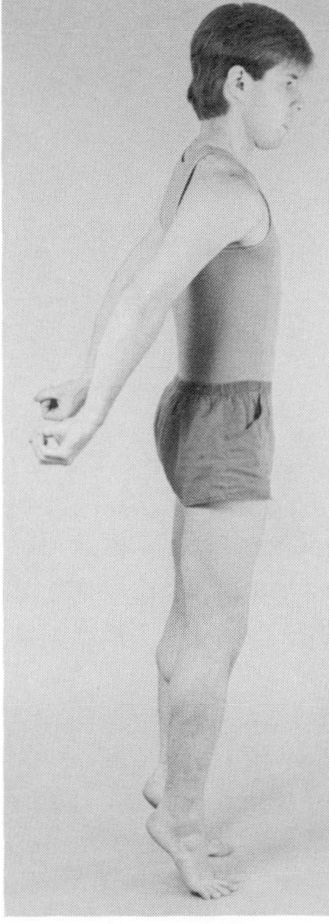

big toes into the floor and look down at a forty-five degree angle. Keep your eyes focused on one point throughout the exercise.

Inhale deeply while slowly bringing the arms backward, fists facing backward, and come up onto the toes. Exhale with a full, loud "hey" sound while forcefully bringing the arms forward, fists forward, and lowering the heels while bending the knees. With the arms and thighs parallel, hold the posture and hold the breath for at least ten seconds. Concentrate all your power in the lower abdomen.

Repeat the entire sequence a few times.

Effects: Brings power to the *hara* or lower abdomen, where the center of gravity should lie.

Relaxation: After performing Yoga poses or corrective exercises, also after working or studying, it is vitally important to relax the whole body and mind. Those who suffer from vision problems obviously retain tension, especially in the neck, shoulders and hands. In these cases, intentional relaxation should be practiced.

The following sequence of movements will enable you to experience relaxation:

Lie on your back and shake the whole body from side to side. Stretch your arms, legs and torso in whatever way and in whichever direction your body tells you it wishes to stretch.

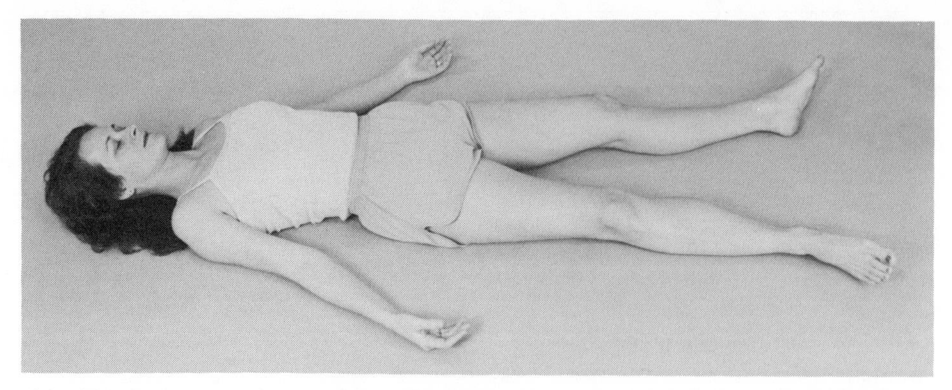

Lie with the legs open shoulder-width and the arms about thirty degrees out from your sides, palms up. Tilt your chin slightly up and let your jaw drop down, opening the mouth. Close your eyes lightly and let each part of your face relax. Imagine your whole body and mind are relaxing, as your breathing becomes deeper and longer. Maintain this posture, the Yoga Corpse Pose, for a few minutes.

Inhaling, raise the hips. Exhaling, lower them. Repeat this raising and lowering a few times. Turn your head slowly from side to side.

Inhale and exhale slowly and quietly. With each inhalation, imagine that your body, mind and eyes are becoming more and more relaxed. See yourself completely relaxed in your mind's eye. Maintain the relaxation for a few minutes.

Rub your palms together. Place them gently over your eyes, letting them rest there for a few minutes.

Eyesight and *Ki* Energy

Yoga comprehends the body as one organism with the functioning of each part to some extent related to the functioning of every other part not only because of the connections interwoven by the nervous and circulatory systems, but also because of a spiritual energy system. Running from the head to the fingertips and to the toes are fourteen *nadi*, the Hindu word for rivers, along which *ki* (electromagnetic energy) flows. These *nadi*, discovered in ancient times, are known as meridians in terms of acupuncture.

Study of the meridians teaches us that the way one uses his hands and arms and feet and legs influences the condition of internal organs and bodily systems. For this reason, Yoga poses and corrective exercises entail manipulations of the arms and legs, enabling otherwise inaccessible parts of the body to be stimulated.

Even each finger, through the meridian connections, relates to other parts of the body, which in turn may affect the eyes or eyesight. The hands as a whole are intimately meshed with the organization of the brain and eyes, which is why the mind and eyes can become relaxed through working with the hands and why the eyes and hands feel strained when the mind tires.

The relationship between the fingers and internal organs or systems is as follows:

Thumb—influences the vagus nerve (a cranial nerve innervating the upper body and most abdominal organs), intestines, liver, and flow of bodily fluids.
Index finger—influences the optic nerve and digestive organs.
Middle finger—influences the circulatory system and intestines.
Ring finger—influences the nervous system, liver, and kidneys.
Little finger—influences the reproductive organs, urinary organs, lungs, and functioning of the sympathetic nervous system in combination with the heart.

These parts of the body can be stimulated by exercising each finger as follows:

- Exhaling, pull each finger outward and rotate it both clockwise and counterclockwise.
- Exhaling, pull each finger backward; exhaling, push each finger forward. Exhaling, pull and push all the fingers forward and backward.
- Exhaling, stretch each finger away from the one to the right of it; exhaling, stretch each finger away from the one to the left of it.

Tsubo Points: Along the body's fourteen meridians, located at places identifiable by sight and touch, lie points called *tsubo*. These points, numbering 365 in all, are essentially loci at which the body's stream of energy, when it becomes sluggish, tends to stagnate.

An acupuncture treatment involves the insertion of a needle at one or more tsubo points in order to balance the energy flow throughout the entire body. Moxabustion, another form of treatment based on the meridian system, stimulates tsubo points by means of heat, while shiatsu massage utilizes finger pressure. All three forms of treatment intend to energize weakened organs while draining energy from overly excited ones.

By massaging yourself you can directly invigorate the eyes or specific parts and functions of the body related to the eyesight. In accordance with the following diagrams, locate each tsubo point indicated. The points will probably feel sensitive. Massage each one by pressing inward with your thumb and holding while exhaling. You may apply as little or as much pressure as is tolerable, gradually increasing the force. Alternate massage techniques such as rubbing, kneading, vibrating, tapping and squeezing are also acceptable forms of tsubo point stimulation.

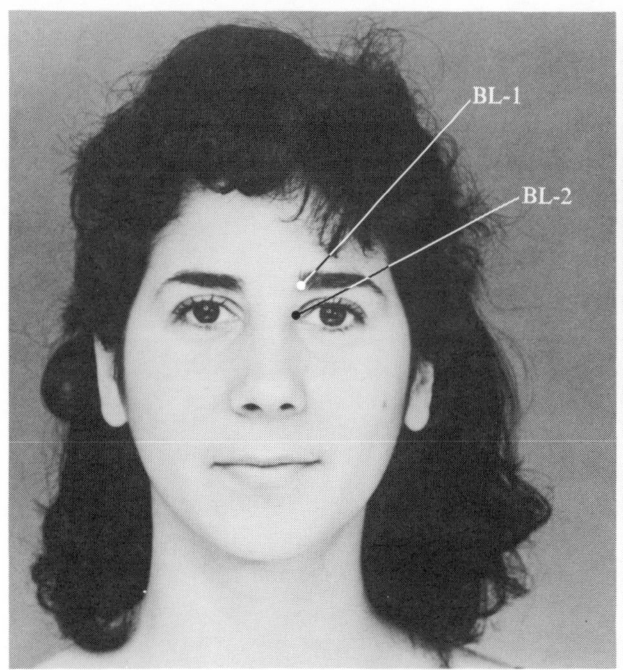
BL-1
BL-2

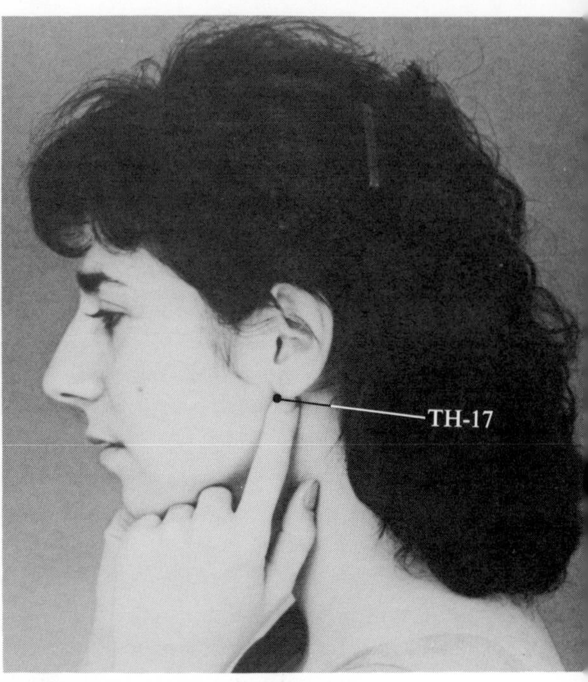
TH-17

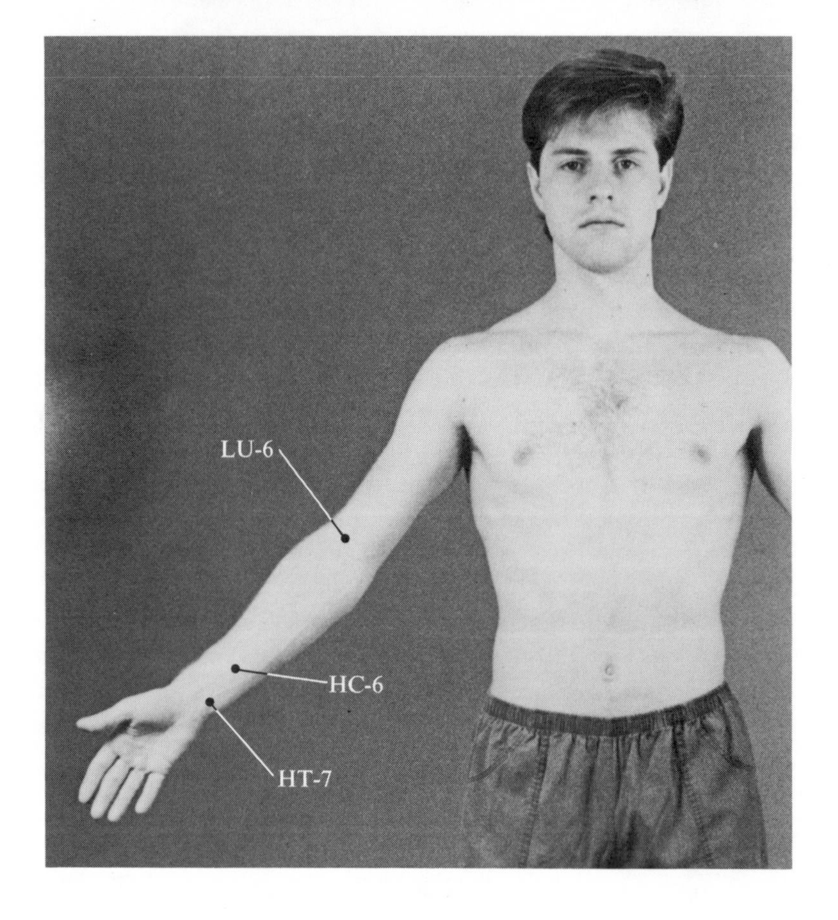
LU-6
HC-6
HT-7

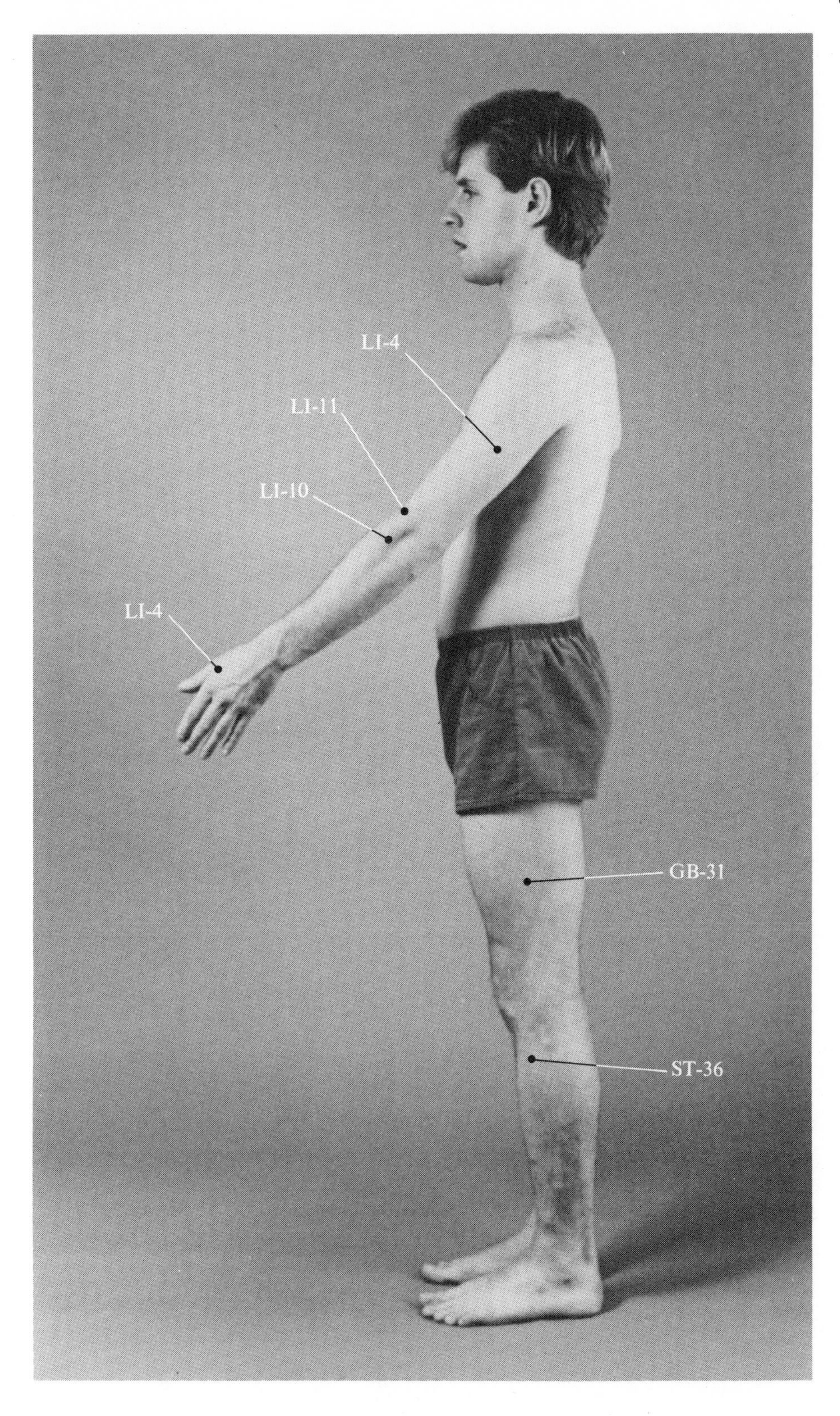
LI-4
LI-11
LI-10
LI-4
GB-31
ST-36

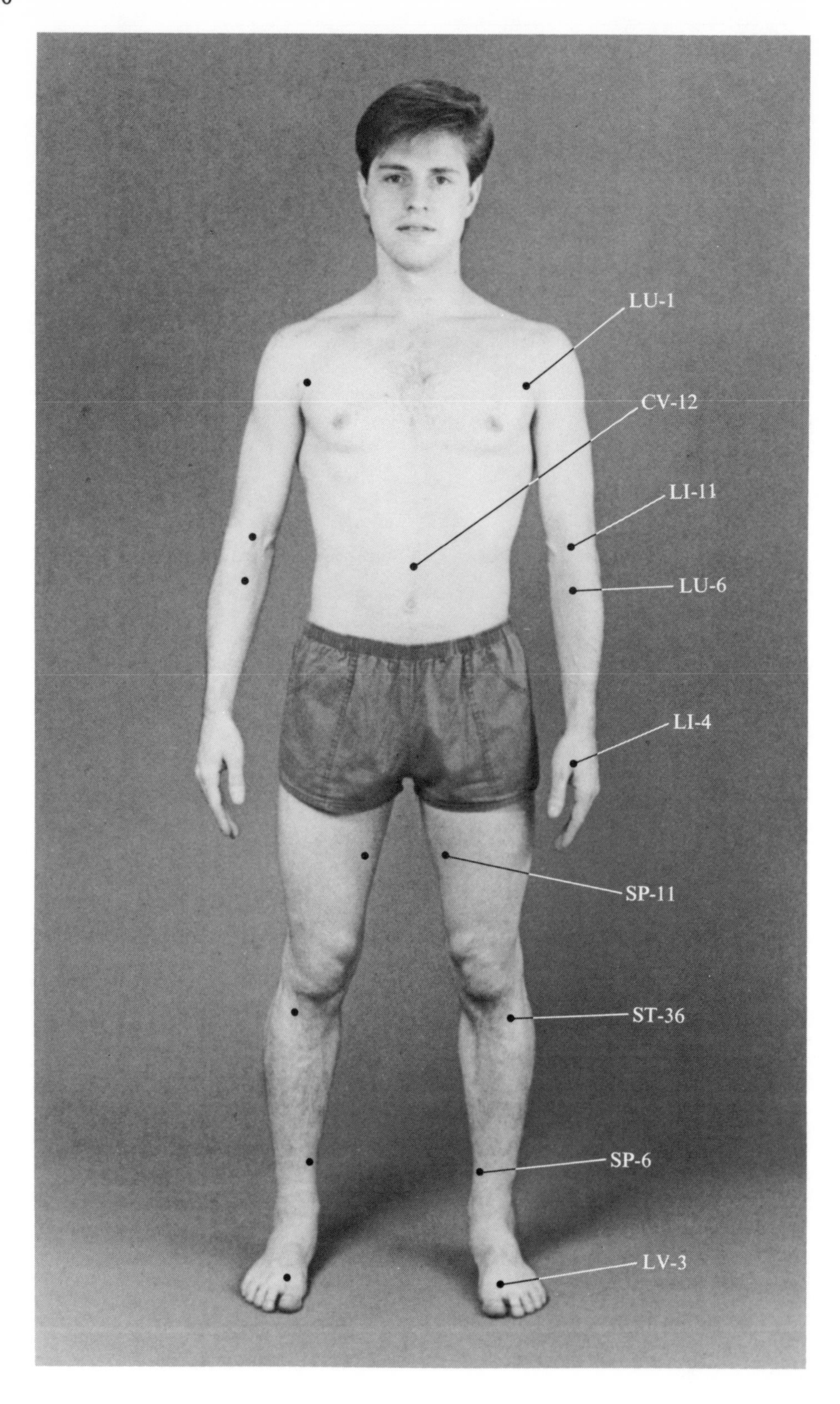
LU-1
CV-12
LI-11
LU-6
LI-4
SP-11
ST-36
SP-6
LV-3

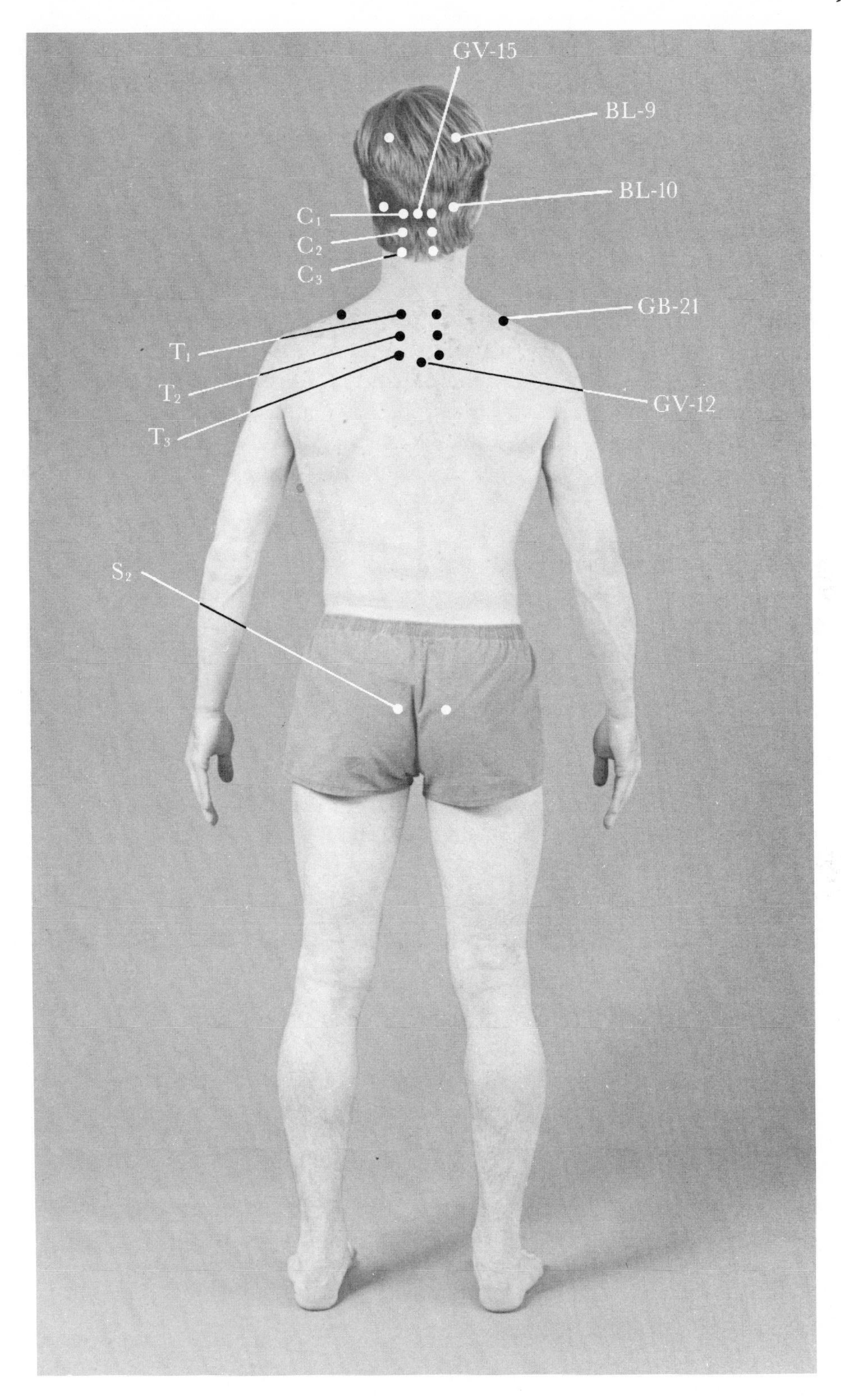
GV-15
BL-9
BL-10
C1
C2
C3
GB-21
T1
T2
T3
GV-12
S2

Running: Another way, besides massage, to activate the flow of electromagnetic energy throughout the body is running. In essence, running is akin to Zazen meditation. Their postures are fundamentally the same, except of course when running one is not sitting. But for both meditation and running the spine is upright, the torso and arms are quiet, the chest is fully expanded, and the shoulders are low and relaxed. Although in running there is propulsion, its energetic source is founded from the waist down, the same area that in Zazen meditation supports and stabilizes the upper body. The manner of breathing, however, for each is different. The best breathing pattern for long distance running is two exhalations followed by two inhalations and repeat.

The ability to run, including the speed and distance, depends upon one's basic constitution and immediate condition. The kind of footstep and degree of arm movement utilized, plus the way of breathing, also influence one's ability to run. To stretch your running capacity, while heightening the body's blood circulation and flow of energy, try the following variations in addition to the standard version of running:

- Run backward, keeping the spine straight.
- Run taking two or three steps alternately from side to side.
- Run by crossing the legs and twisting the upper body.
- Run on the toes only.
- Run lifting the knees up high.
- Run trying to kick the buttocks with the heels.
- Run taking long strides.
- Run while exercising the arms in all directions.
- Run bringing the shoulder blades together for twenty steps and for the next twenty steps bending the head side to side, and repeat.

Eyesight and Breathing

The ability to see clearly significantly depends upon the power of breathing. If one's inhalation is shallow and exhalation short, then the elimination of gaseous toxins is surely incomplete and the blood impure. As a result, the nervous system will be weakened, with possible debilitation of the optic nerve and paralysis of the eye muscles.

Taking in plenty of oxygen and exhaling completely strengthens the nervous system and cleanses the blood. Oxygen enables the muscles to relax and food to turn into energy. The Yoga discipline of *pranayama*, or breathing exercise, is a means to increase the intake of oxygen and expulsion of carbon dioxide, while improving the blood circulation overall. And insofar as one's breathing is deep and long, the eyes will not tire easily.

Guidelines: The following guidelines are to be incorporated into the performance of all Yoga breathing exercises:

- Be aware of using the whole body since the whole body is affected by respiration.
- Inhale deeply and exhale fully, making the exhalation longer than the inhalation.
- Breathe in from the lower abdomen up into your eyes and breathe out through your eyes.
- When an exercise calls for *kumbaku* (holding the breath), retain the breath for approximately ten seconds in the beginning of your practice; later on, gradually increase the time of holding.

Breathing Exercise: The breathing exercises below might best be practiced in the morning and may be done in conjunction with the Dynamic Breathing Exercise (beginning on page 100). They should not be practiced until one to two hours after eating. For all of them, sit in the Seiza Posture (as illustrated below and on page 134) or sit on the edge of a chair. In either posture, keep the spine upright and the eyes closed or relaxed.

1. Deep Breathing Exercise—To relax the body and mind

Place the fingers under the base of the rib cage. Inhaling through the nose, bring the shoulder blades together behind you and arch backward. Exhaling through the mouth, press the fingers up under the rib cage and bend forward, completing the exhalation. Repeat the entire exercise a few times.

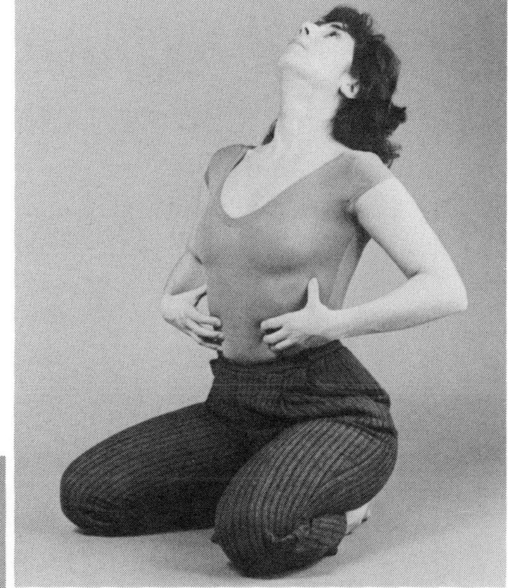

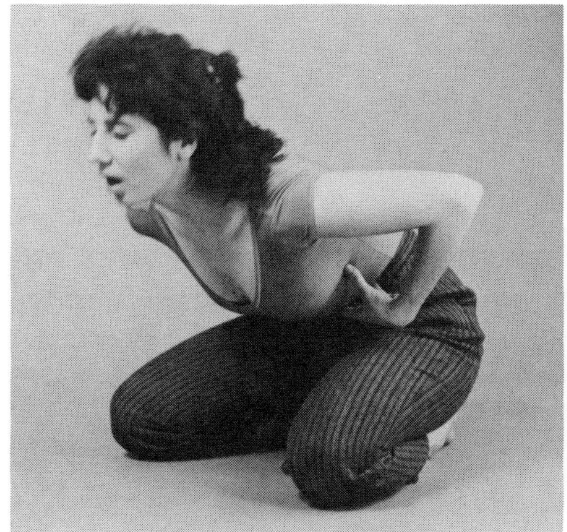

2. Fundamental Breathing Exercise—To increase the breath capacity
Exhale through the mouth, pulling the abdomen in. Relax the abdomen. Inhale through the nose, pushing the abdomen out, opening the chest, and raising the shoulders. Exhaling slightly, lower the shoulders, close the anus and hold your breath in the *hara* (lower abdomen) for approximately ten seconds.

Continuing the exhalation, pull the abdomen in. When the exhalation is nearly complete, relax the abdomen and stretch the spine upward. Inhale naturally. Practice the exercise over and over for from five to ten minutes, gradually increasing the length of time you hold your breath.

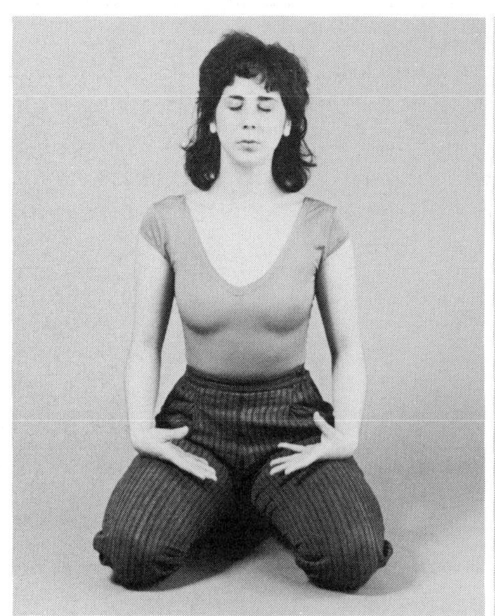

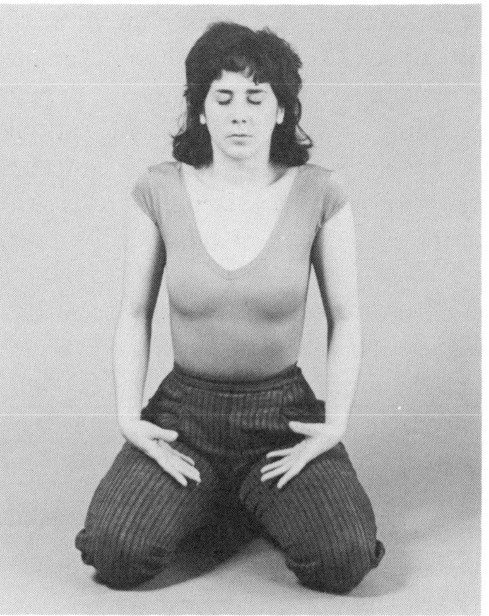

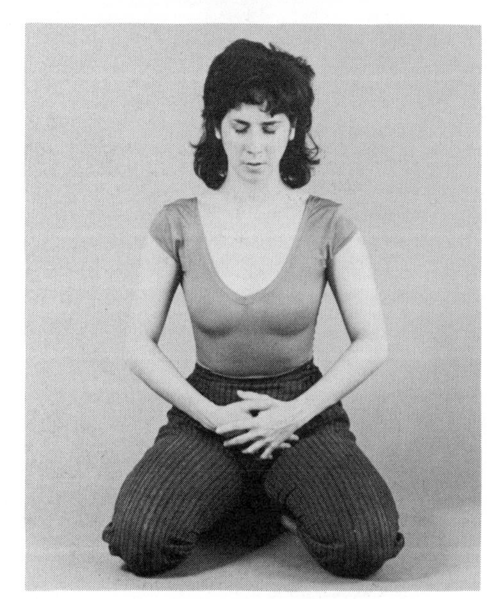

3. Concentration Breathing Exercise—To integrate the body and mind
Inhale through the nose, up from the lower abdomen into the chest, opening the chest fully. Exhale, making a "humph" sound that reverberates up from the *hara* (lower abdomen), and close the anus. Hold your breath and concentrate your power in the *hara*. Slowly and steadily exhale.

Slowly and steadily exhale to completion and relax. Repeat the entire exercise a few times.

4. Stability Breathing Exercise—To stabilize the breath
Open the knees wide. Make a fist of each hand with the thumb inside. Stretch the arms straight out to the sides at shoulder height. Bend the elbows to a ninety degree angle with the fists pointing upward. Turn the wrists backward, palms up.

Inhaling through the nose, arch the head backward and bring the shoulder blades together behind you, while rolling your eyes upward. Exhaling through the mouth and keeping the elbows at shoulder height, turn the forearms parallel to the floor and bend the head and wrists forward.

Continuing the exhalation, stretch the arms straight out (onto the floor if in the Seiza Posture) in front of you, stretching the torso forward until the exhalation is complete. Repeat the entire exercise a few times.

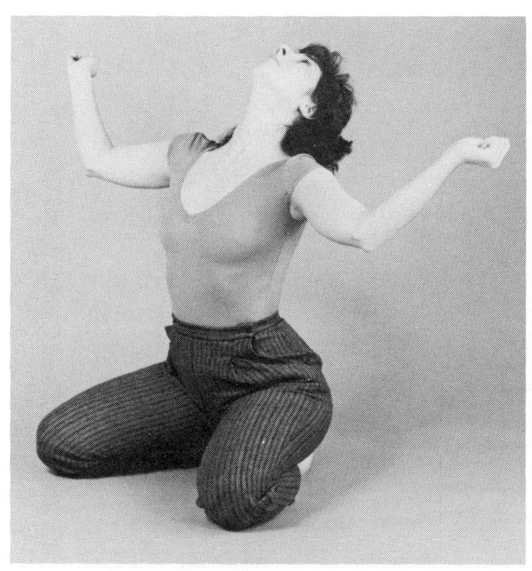

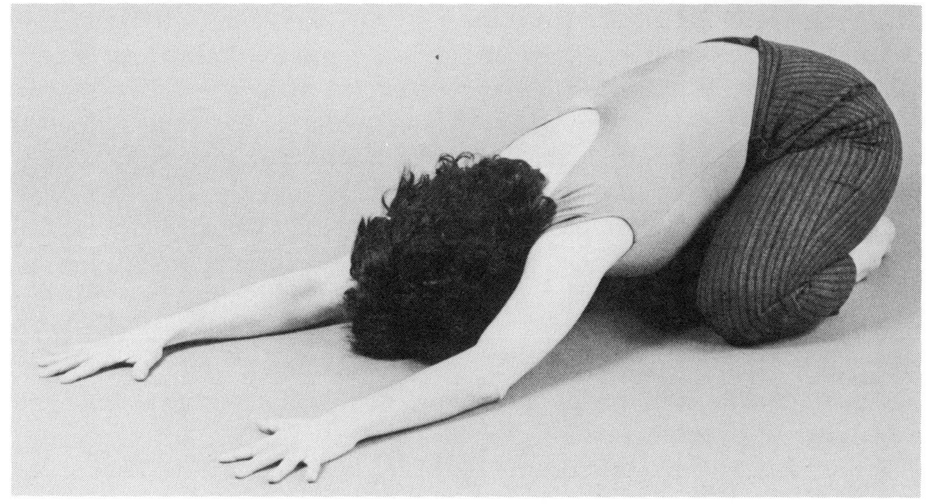

Eyesight and Diet

In contemporary society, people's eating habits are becoming more and more unnatural. So are the methods of food preparation. Food itself is often artificially processed and, in general, far removed from its original state. As a result, the standard diet of modern times lacks variety and the average appetite is excessively specific. It could be said that most adults, conditioned to eating processed, canned, frozen, chemicalized, dyed and unnatural foods since childhood, have lost their senses of taste and smell.

The natural way of food selection, as with animals, is to know instinctively which foods are good and which are harmful for you. To do so, you must recover your natural appetite (perhaps through fasting), so that you can select food fit for you as a human being and your individual needs. If your food is appropriate, it will be as medicine that helps you. If not, it will be as poison.

In the scientific concept of nutrition, individual needs are not taken into consideration. Food is merely analyzed according to its component vitamin and mineral elements and calories are measured. Food is then categorized, and the criteria for selection are more quantitative than qualitative. Eating this way, according to a theory of nutrition, is actually non-nutritious.

The Yoga outlook on food is based on observance of the principles of natural law with consideration for individual needs. Food itself, of course, is neither good nor bad. But an understanding of natural law tells us that the best foods for a human being are those which can be eaten whole. That is because whole foods, including whole grains and the roots plus the leaves of vegetables, as well as food from the earth and from the sea, enable the blood to maintain a normal acid/alkaline balance.

Other propositions to be learned from natural law regarding food are to eat only what is capable of growing within one's climatic zone; and to eat only foods that are in season. Honoring these, the body can harmonize with Nature.

For example, natural intuition leads us in the spring to desire leafy green vegetables. Since greens promote bowel movement, the plants are Nature's provision for cleansing the body of heavy foods consumed during the winter. In summer, people naturally crave more acidic and sour foods, foods such as cucumber, eggplant and watermelon, which cool the body and balance the alkalizing effect of heat. In the fall, root vegetables and oily foods such as fish are desirable to prepare the body for the cold of winter. In winter, a natural appetite prefers dried food, such as roots and sea vegetables, which have stored within them the warmth of the sun and the salts and minerals of the earth. If we then maintain a variety of grains as the staple food, this way of eating, in accordance with natural law, empowers us to adjust to the environment and its vicissitudes. The ability to adjust is the mark of a healthy person.

If you neglect the requirements of the life-force and ignore the laws of Nature;

if you simply count calories, measure the components of food, or weigh proteins against carbohydrates; if you disregard whether your food is whole or where it has been grown or if it is artificially processed; even if you are completely vegetarian but eat food out of season or eat raw foods in winter or eat tropical foods in a temperate climate—then, the relationship between your body and mind will become unbalanced. That is because your way of eating is out of kilter with the environment and its provisions, as if at war with Nature. Imbalance between the body and mind is one of the major sources of sickness.

The natural way of eating, then, is to choose, from all variety of food grown seasonally within one's climatic zone, the highest quality of the food best for oneself; and to find the appropriate ways of cooking or preparing those best foods.

The food served at the Zen Yoga Dojo in Japan is basically whole, including whole grains, land vegetables, sea vegetables, and fishes with the head, bones and skin intact, small enough to be eaten whole. Raw, pickled and dried foods, including dried *tofu*, mushrooms, *daikon* (large white radish), fish and squid, are also provided seasonally. Again depending upon the season, vegetables are prepared raw, steamed, boiled or sautéd. *Miso* soup, containing land and sea vegetables, is the daily breakfast. Sea vegetables such as *kombu*, *hijiki*, *arame* or *nori* are served at dinner or supper. In other words, each day different foods are prepared or the same foods are prepared in different ways. The main meal, the dinner meal, is served at around one in the afternoon. The quantity of food alloted each person is comparatively small, but the variety, based in myriad food selection and diversified food preparation, is wide. As a result, a day's meals are rich in an assortment of vitamins, minerals and colors.

Since healthy eyes require plenty of oxygen plus vitamins and minerals, nutritionally unbalanced people are likely to become nearsighted. For example, if the body is deficient in calcium, the white of the eye or sclerotic, supposed to uphold the form of the eyeball, loses flexibility. If one then reads or does paperwork for prolonged periods, the eyeball will expand and stay so. People attracted to sugar and other acid-producing foods are highly subject to eye troubles due to unbalanced blood.

What is good for the liver and kidneys is good for the eyes. A therapeutic diet and Chinese herbs help the nervous system relax and build up the liver and kidneys. Both these organs need an abundance of vitamins and minerals; and since both greatly affect the eyesight, the liver and kidneys should be strengthened through naturally nutritious food and the practice of Yoga poses.

The vitamins and minerals comprising a naturally nutritious diet actually work supportively to enhance each other and the life-force. For example, brown rice, brown barley and other whole grains, containing the germ or vitamin components of food, in combination with sea vegetables and raw and cooked vegetables containing minerals, are mutually augmenting. If you pickle vegetables in *miso*, their content of vegetable protein and vitamins A, B and D will be increased; and if you include hot peppers in your pickle jar, the vitamin C content of the other vegetables will be increased.

Pickles in general, that is, food pickled in sea salt, *miso*, or bran, rather than commercially pickled in vinegar, are excellent sources of vitamins and minerals. Pickled fish and sea vegetables provide protein, calcium and iodine. Also, pickles serve to revive one's senses of taste and smell, so necessary for the recovery of the natural appetite.

The quantity of food taken and the manner of eating it also affect the eyesight. The Oki Yoga Dojo in Japan recommends fasting for people who wish to improve their vision. Even if a person fasts only, and practices no other disciplines to heal the eyes, his vision will improve to some degree. Through fasting, all the muscles of the body relax, the nervous system relaxes, the blood becomes clean, the purification mechanism of the body is strengthened, and the eyes can see much more clearly.

To sustain or to heal your vision, whether before or after fasting, indeed at all times, it is necessary to refrain from overeating and to chew well. These practices establish a normal acid/alkaline blood balance and aid in complete evacuation. The foods you choose and your methods of preparing them will contribute markedly to whether you see well near and far, day and night, or not. In other words, if you consistently provide yourself with stimulations good for you, you will regain and maintain your ability to see well.

Chapter 5

Healing the Vision

Even if one person's illness is called by the same name as another's, even if both people exhibit what appear to be identical symptoms, Yoga counsels looking for idiosyncratic causes for each case of disease and its symptoms. Because Yoga insists upon individuality, it offers a comprehensive approach to self-healing, including methods for correcting the breathing, posture, diet, mentality, spirituality; in short, the life-style. Nevertheless, the aim of all Yoga healing methods is the same; that is, to integrate the mind, body and spirit for the one purpose of increasing personal human potential. Even Yoga poses, which on first acquaintance might seem to be but a form of calisthentics, intend to make the body cooperate in the practitioner's spiritual development. Yoga poses therefore transcend other exercise; indeed, at the Oki Yoga Dojo in Japan, poses and corrective exercises are designated as *dozen* or Zen in motion.

Through Yoga, you can discover your own best method or methods of self-healing. By accepting new stimulation, you can determine what you need to correct your vision. Yoga exercise, as a form of new stimulation, is an essential part of the healing process. This book presents a variety of exercises for changing oneself, and through that for improving the eyesight. Through experimentation, you can find those exercises that refresh you, energize your eyes, and improve your vision. However difficult the exercises which reap these benefits might be for you to perform, they are the exercises appropriate for you.

After several weeks of practicing Yoga exercise, after you begin to see and feel better in general, you will reach a plateau in the healing process. But do not stop there. Continue on with the exercises you have selected, advancing your execution of them. Eventually you will reach another upward stage, which, in turn, will level off for a time. In other words, any exercise that is good for you need not ever be abandoned entirely. Your form within it can always be brought nearer to perfect balance, and variations can be devised to render it more challenging for your body if it has become easy to do.

It should be noted that, in general, corrective exercises are not supposed to be easy. If a Zen Yoga corrective exercise is easy for you, it is probably unnecessary and relatively ineffective. You are advised to base your selection of exercises on how your body and mind and eyes feel after doing the exercise rather than on how well you do it. Concentrate, therefore, on the process of execution more so than on completing the exercise. For example, if a corrective exercise calls for a sit-up, the attempt to sit up rather than the achievement of sitting up is what produces the correction.

By the look of some exercises, you might think they are not affecting the eyesight. But through the connection of the eye nerves to the entire nervous system; through the relationship of the eyes to internal organs; and through the stimulating relationship between arm and leg use and eye use, all the exercises outlined do affect the vision.

Dynamic Breathing Exercise

Dynamic breathing exercises consist of combining intentional movement with conscious breathing. Their function is to increase the opening and closing capacity of the rib cage and the strength of the muscles involved in respiration. Their aim is deepened breathing.

The following dynamic breathing exercises are recommended for practice in conjunction with a program of corrective exercise. For example, the breathing exercises (followed by the purification exercises outlined below) might be performed in the morning, while corrective exercises are most effective if performed before retiring for the night. Of course, the breathing, purification and corrective exercises can all be attempted in a single session. Each breathing exercise may be repeated in its entirety several times.

1. Stand with the feet parallel, open shoulder-width. Relax the shoulders, let your arms and hands hang relaxed at your sides, relax your face, and look straight ahead.

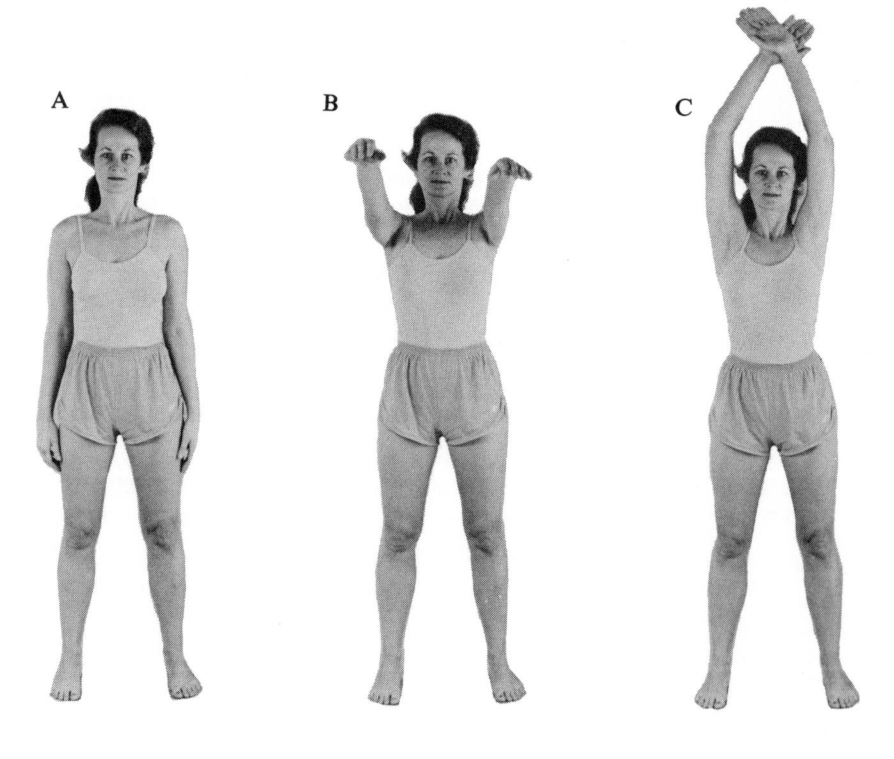

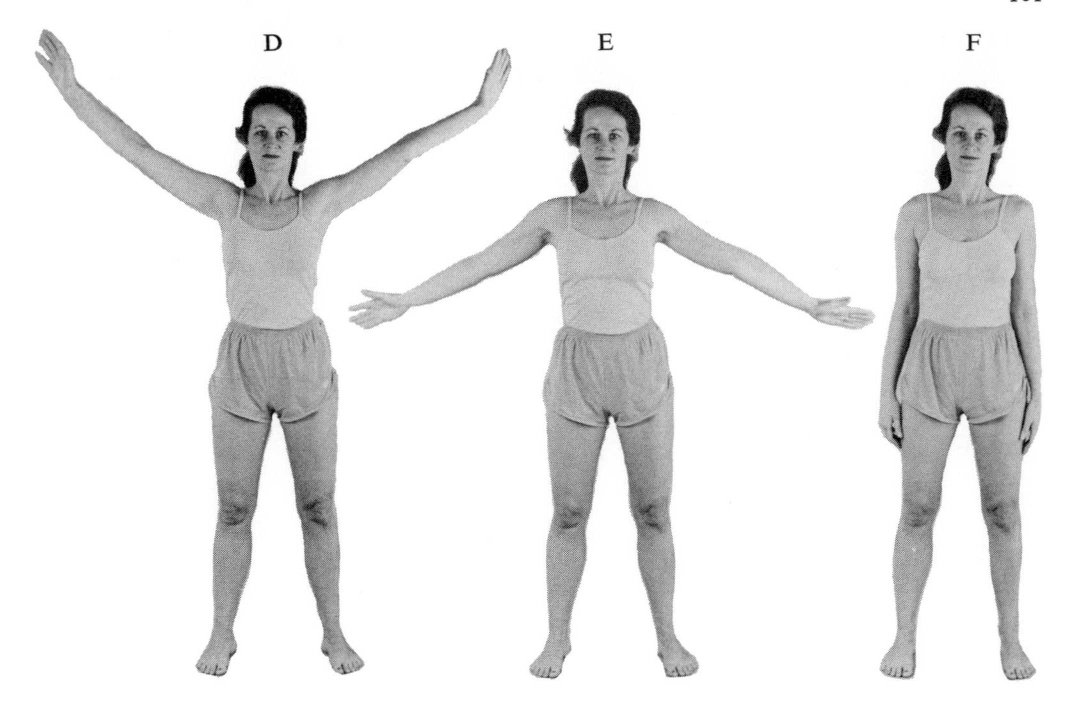

Inhaling through the nose, slowly and naturally raise the arms, keeping the hands relaxed, up over the head (A–C). Cross one hand over the other, close the anus, and hold the breath.

Exhaling through the mouth, slowly and naturally lower the arms out to the sides, palms down (D and E). Completing the exhalation, relax in the original posture (F).

2. Stand with the legs open twice pelvis-width. Expand the rib cage and stretch the arms straight out to the sides.

Inhale slowly and deeply through the nose. Exhaling through the mouth, slide the left hand up to the armpit, stretching the left side of the torso. At the same time, bending to the right, slide the right hand down the right leg. Look at the right foot.

Inhale slowly and deeply through the nose and return to center. Exhaling, perform the movement on the opposite side.

3. Stand with the legs open slightly. Cross the wrists in front of the body at the pubic bone, close your eyes, and relax.

Inhaling from the lower abdomen, open your eyes and take one step forward onto the right foot, bending the right knee and keeping the left leg stretched. At the same time, stretch the arms out sideways and twist them outward. Focus your eyes on a point ahead.

Exhaling, return to the original posture. Inhale and perform the movement on the opposite side.

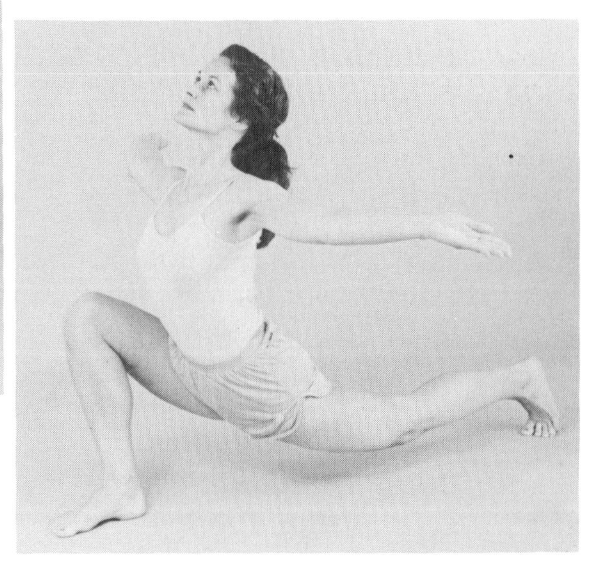

4. Stand with the legs open slightly. Cross the wrists in front of the body at the pubic bone, close your eyes, and relax.

Inhaling from the lower abdomen, open your eyes and take one step forward onto the right foot, placing your weight there, while raising the left heel and flexing the left toes onto the floor. At the same time, bend your elbows, trying to bring them together behind you, and elevate your chin. Roll your eyes upward.

Exhaling, return to the original position. Inhale and perform the movement on the opposite side.

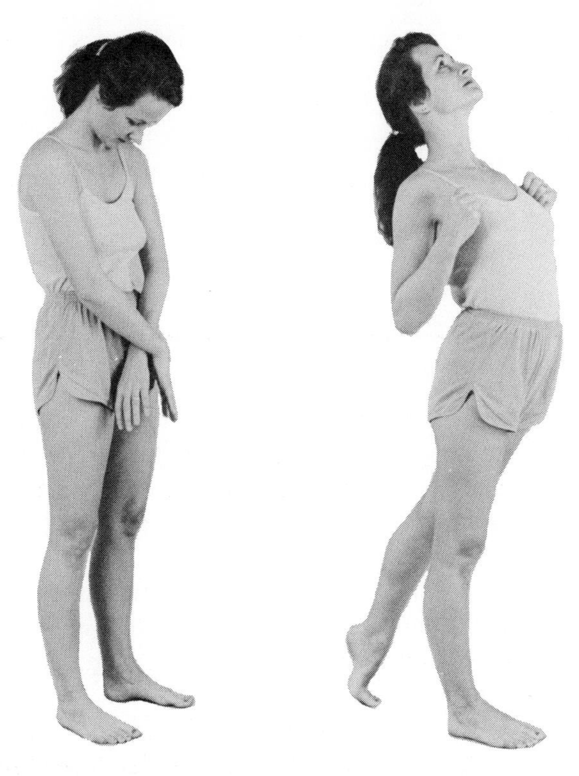

5. Stand with the legs open pelvis-width.

Inhale slowly and deeply through the nose. Exhaling through the mouth, bend forward from the waist, stretching the arms backward and upward while twisting them to the inside. Pull the abdomen in and close your eyes. Inhaling from the lower abdomen, come up

from the waist, putting your weight on the right foot, stretching your left leg backward, heel lifted and toes pointed into the floor, arch the torso backward and stretch the arms upward and backward over the head while twisting them to the outside. Open your eyes and look at your hands.

Exhaling, return to the original posture. Inhale and perform the movement so the leg positions are reversed.

Purification Exercise

Purification exercises consist of movement to increase the opening and closing capacity of the rib cage (to increase the breathing capacity) and of the pelvis (to stimulate blood circulation to the lumbar and abdominal regions). Their aim is improved elimination and the discharge of old waste material impacted in the intestines, as well as purification of the mind.

The following purification exercises are recommended for practice in the morning, just after the dynamic breathing exercises. While breathing exercise increases the flexibility of the upper body muscles, purification exercise makes peristalsis of the lower body more fluent.

For all the purification exercises, concentrate on the *hara* (lower abdomen).

1. Stand with the legs open shoulder-width and the feet parallel. Stretch the arms straight out in front of you and interlock the fingers, palms facing outward.

Inhale deeply and hold the breath in the lower abdomen. Exhaling, move the

arms up and down until the exhalation is completed. Inhale and rest a moment.

Repeat the exercise several times.

2. Stand with the legs open and the feet parallel. Stretch the arms straight out in front of you and interlock the fingers, palms facing outward.

Inhale, bringing the backs of the hands straight into the chest. Exhaling, stretch the torso and arms straight up. Inhaling, return the backs of the hands to the chest. Exhaling anew with each stretch, continue the stretching movement at various angles (for example, at the two, four, six, eight and ten o'clock positions)

all around in a big circle. Between each stretch inhale and return the hands to the chest. Follow the hand movement with your eyes throughout.

Repeat the circle several times both clockwise and counterclockwise.

3. Stand with the legs wide open and toes turned inward. Place the hands at shoulder-width on the floor in front of you and spread the fingers. Keeping the arms and legs stretched, inhale and push the hips upward and backward, giving the spine a good stretch. Holding the posture and the breath in the lower abdomen, relax your eyes.

Exhaling powerfully, open the rib cage and lower the pelvis to the floor. At the same time, arch the chest and head upward and backward. Keep the arms stretched and follow the head movement with your eyes throughout.

Repeat the hips-up movement slowly on an inhalation and then the pelvis-down movement quickly and vigorously on an exhalation from ten to fifteen times.

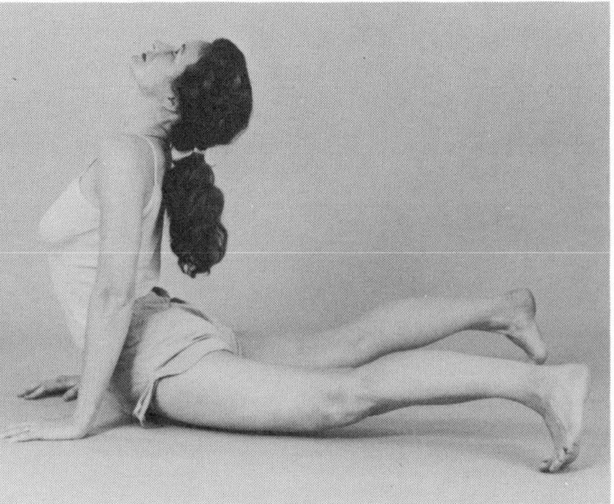

4. Squat down with the feet pelvis-width, heels up, and hands down. Relax the upper body and inhale deeply, holding the breath in the *hara* (lower abdomen).

Exhaling strongly, kick the legs, stretching them out behind you onto the toes, while arching the torso and head upward and backward. Push the pelvis into the floor, open the chest, and look at the ceiling while holding the posture.

Return to the original position and repeat the exercise several times.

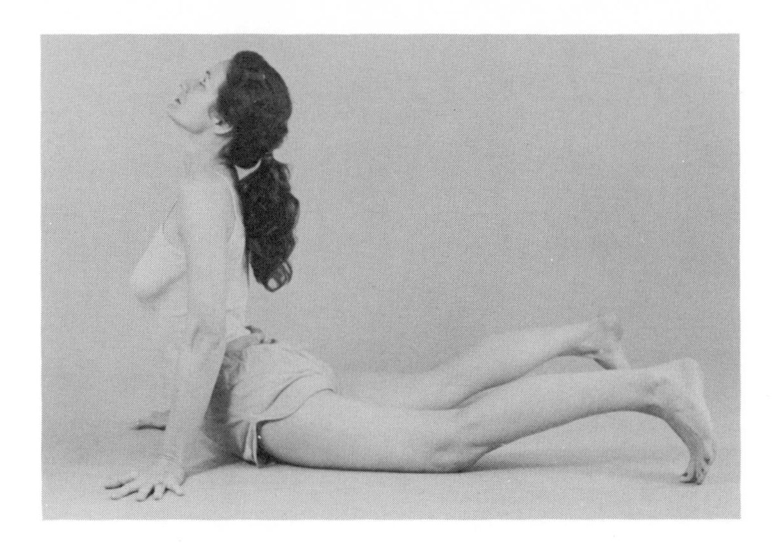

5. Lie facedown with the legs wide open and the toes pointed into the floor. Place your hands at the sides of the chest, fingers pointing forward.

Inhaling, arch the torso and head upward, straightening the arms and pressing the pelvis into the floor. Exhaling strongly, twist the whole body to the left and

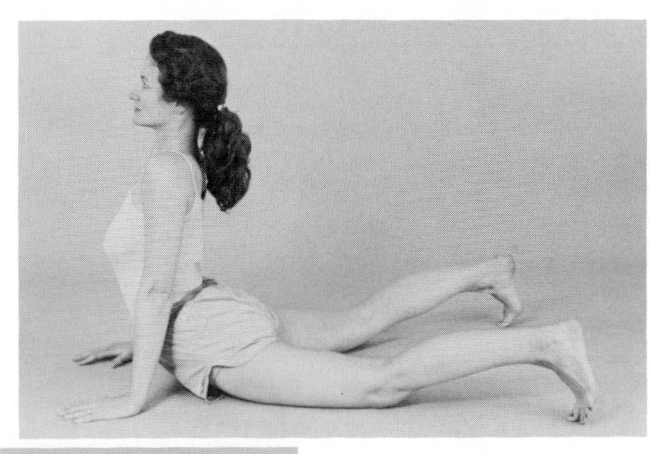

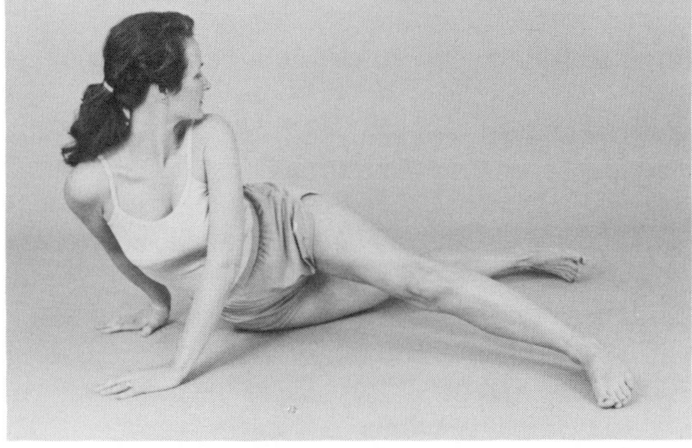

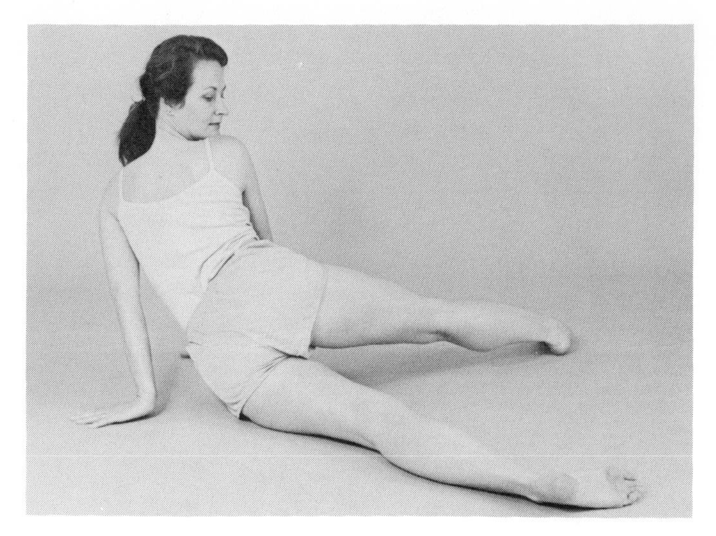

look at the right heel. Inhale and return to center. Exhale and twist to the right, looking at the left heel.

Repeat the twisting from side to side ten to twenty times.

Corrective Exercise Theory

The form of healing most free of artifice or technique is that accomplished while sleeping. A healthy person moves about in his sleep, subconsciously correcting imbalances of the posture and internal organs. This movement is corrective exercise at its most natural stage.

Healing yourself while sleeping is possible to the extent that you are able to sleep deeply, which depends upon being able to relax when awake. Relaxation, which is unequivocally the natural state of a human being, might be defined as a condition free of overexcitement or overstimulation. It is a state in which the mind is positive, so that experiences at hand are accepted without undue mental agitation and physical resistance. If the mind and body become tense, or the blood circulation poor, or if there is organ imbalance, it becomes impossible to heal oneself while sleeping. Tension, the mark of an inability to heal oneself unconsciously, is the herald of chronic disease. Teaching the body and mind to relax is the key to recovering one's native, unconscious healing power.

There are various methods to promote relaxation while awake so as to deepen the sleep. Conscious corrective exercise, intentionally practiced, is a method particularly appropriate for those who have lost their ability to adjust themselves while sleeping.

The primary purpose of conscious corrective exercise is to recover a natural way of breathing. When the breathing becomes natural, you can sleep well. Corrective exercise also intends to balance the posture and the functioning of the nervous and hormonal systems, and to aid excretion and elimination. By means of conscious corrective exercise, the whole body can be changed and the mind made more stable. Withal, the five senses, including the eyesight, can strive for an acceptable sensitivity.

Corrective Exercise and Breathing: Yoga teaches us that one's breathing is a manifest representation of his body and mind. A person's overall condition can be diagnosed through observing his way of breathing. Unbalanced people, people with mental and physical ailments, breathe unnaturally. Specifically, certain sicknesses produce certain types of breathing. Also, before all else changes, before any segment or aspect of the body and mind changes, the breathing changes. The key, therefore, to controlling the body and mind is control of the breathing. By correcting your breathing, by normalizing it, you affect your sickness, or, rather, your health.

One's way of breathing even shows up in the eyes. For example, when the mind and body become tense through prolonged concentration, that is, when the upper body becomes unconsciously tight, the breathing will become shallow and short or even stop periodically. As soon as the breathing stops, the eyes stop moving too. One becomes tired, even without noticing it. And if concentration is sustained without moving for a long time, if the halting breathing persists, the eyes, also tired, will become near-permanently fixed.

As has been stressed, to secure optimum vision, three factors must be present: 1) a relaxed body and mind; 2) good blood circulation; and 3) clean blood. Behind all three is a normal way of breathing.

If you can breathe in deeply, your body and mind will become calm. But if the inhalation is short, if the breathing stops and starts, the body and mind will become tense. If you can exhale lengthily, as happens when one is laughing, the body and mind will relax. When using your body and mind, practice a kind of breathing as if laughing. When using your eyes, breathe from them, exhaling through them. Then your neck will relax, and your eyes, along with your breathing, will begin to move naturally. Deep, stable, long breathing naturally relaxes the body and mind, increases blood circulation, and, by stepping up the intake of oxygen, purifies the blood, giving vitality. Under these conditions, the native healing power works best.

The reason that poor posture harms the eyesight is that poor posture disables the breathing, making it shallow. The reason that mental tension harms the eyesight is because it, too, shortens the breathing. One characteristic of Yoga poses is an emphasis upon breathing, while Zazen practice emphasizes correct posture plus breathing. Yoga and Zen together teach us to rid ourselves of chronic disease by correcting the breathing.

Conscious corrective exercise, based on the teachings of Yoga and Zen, was created by Masahiro Oki expressly to enable the individual to correct his way of breathing. If you practice corrective exercise, your shallow, or short, or irregular breathing will change to normal breathing.

Corrective Exercise and Movement: All corrective exercises, indeed all Yoga

poses, are based upon some combination of the following twelve possible ways of moving: bending forward and bending backward; bending to the right and bending to the left; twisting to the right and twisting to the left; standing up and stooping down; contracting muscles and expanding muscles; stretching and relaxing.
A human being, because he can perform all of these twelve movements, is the most flexible of animals, even more so than a cat. Though a cat's body is exceedingly soft and pliant, it cannot bring its shoulder blades together nor bend backward, while a human being can.

According to the kind of movement one is performing, the stimulation given the body differs. Contracting movement, such as a forward bend (which closes the rib cage and chest muscles), stimulates the sympathetic nervous system, making the blood more acidic. Expanding movement, such as a backward arch (which opens the rib cage and chest muscles), stimulates the parasympathetic nervous system, making the blood more alkaline. Yoga practice aims to combine opposite though complementary movements.

The same principle can be seen to operate in respect of movements of the head. When trying to remember something, one generally elevates the chin, thereby stimulating the memory part of the brain. When thinking, the chin is generally dropped forward, stimulating the cerebral cortex. When trying to correlate two concepts, the head is generally dropped forward and alternately side to side, stimulating the brain to put one plus one together.

Variations in eye movement also provide different stimulations. When both eyes look to the right, the eye moving toward the inside (left eye) receives more acidic stimulation, while the eye moving toward the outside (right eye) receives more alkaline stimulation.

Eye movements likewise influence other parts of the body. When the eyes are centered straight ahead, the spine is stimulated. When the eyes are turned downward, the abdomen is stimulated. When the eyes are turned upward, the brain is stimulated. In general, in daily life, the eyes move side to side and up and down. But when one becomes sleepy, the eyeballs protrude, giving a relaxing stimulation. When the stability of the posture is heightened, as in meditation, the eyeballs recede, giving a contracting stimulation.

Several of the corrective exercises presented below include eye movements. And with all the Individual Corrective Exercises specific benefits are denoted. The overall purpose of corrective exercise, however, as stated, is to change the breathing, so that the body and mind might relax sufficiently to heal themselves. Eye movements added to corrective exercises are incentives to the life-force to correct the vision.

Corrective Exercise and the Eyes: One's posture very much influences the disposition of the body and mind, the eyes not withstanding. For example, if the neck is bent forward habitually, blood will congest in the head, increasing the interocular pressure. The eyeballs will expand, causing the ciliary muscles to lose flexibility. The resulting condition is pseudomyopia. When the whole posture is markedly unbalanced, that is, when distinct differences can be noted between the right and left sides of the body, astigmatism or presbyopia might emerge.

Postural imbalances, leading to eyesight imbalances, can be rectified through corrective exercise. Certain corrective exercises, consisting of opening, stretching and twisting movements, affect the vision directly. The nervous system is activated through manipulation of the spine and arm and leg muscles. By correcting spinal and arm and leg imbalances, tension is released from the eyes and the eye muscles can relax. At the same time, blood circulation to the upper body and head is improved.

For example, sit with the legs and arms outstretched in front of you. Stretch the Achilles' tendons. Make fists and twist the arms both inward and outward. (Twisting the arms contracts the joints, thereby stimulating tsubo points along the arms related to the eyes.) Stretching the arms, gradually open them out to the sides. At one angle in particular, your eyes will let go of tension. (Stretching the arms relaxes the muscles, thereby tonifying tsubo points along the arms related to the eyes.) It is through the twisting and stretching that the blood circulation improves throughout the body and to the eyes.

By practicing corrective exercise, you can pinpoint your physical imbalances and seats of physical tension. This is a basic principle and benefit of corrective exercise —that is, in doing corrective exercise you can examine the power of your right side compared to the power of your left side, and you can examine the power of areas of your spine, determining exactly where the physical sources of myopia or vision weakness lie within your body.

Corrective Exercise Practice

Corrective Exercise is best practiced before going to bed at night. Afterward, meditation is strongly recommended to stabilize and hold the relaxation attained. With these disciplines, not only will you sleep better, but upon awakening you will feel revitalized.

Since the major cause of nearsightedness is mental and physical tension, do muster up the courage to relax yourself consciously. Acupuncture and shiatsu massage can aid in relaxation. But by yourself, through deep breathing, through a positive mental attitude, and through acid/alkaline balanced nutrition, that is, through practicing Yoga in an integrated way, you can assure the revival of your self-healing power.

Preparation: The following are guidelines to be observed for the performance of corrective exercise.

Before exercising, wait at least an hour after eating, two to three hours if you eat animal or heavy food, so that your meal is digested. Go to the toilet.

Before beginning, activate your *ki*-energy flow by self-massage of tsubo points, warming-up movement, and breathing exercise. Then intentionally yawn several times.

Make the experience of corrective exercise enjoyable for yourself. If you perform grudgingly, the body will remain tense and its center of gravity unstable.

Assume a state of *mu shin* (empty mind) which is the mind of meditation or non-attachment. Rather than wishing to be healed through the exercises, rather than attaching to a potential result, simply execute the movements in obedience to the needs of the body. Leave the mind free of expectations.

Check to see which of your arms and which of your legs is shorter. Then, when performing those exercises that call for repeating the sequence of movements twice as many times on the more difficult side, consciously stretch the shorter arm and the shorter leg during the repetitions. This conscious stretching helps create an overall balance.

Pay close attention to your breathing. Try to inhale deeply and fully, first filling the lower abdomen and then letting the chest expand. Make the exhalation long and strong, longer than the inhalation, and in the opposite direction, from the upper body down to the lower abdomen. In general, the most difficult movement of an exercise, that is, the movement geared toward eliciting change, should be done on an exhalation. When exhaling, the body releases tension. Fundamentally, combine movement with breathing in the most natural way.

At the start of each exercise, concentrate on the *tanden*, that spiritual point within the very center of the lower abdomen about three fingers below the navel, deep inside between the front and the back. Then visualize the movement you are about to execute and, sending your power out from the *tanden*, perform the movement.

Generally, keep your eyes open throughout each exercise, but move your eyeballs behind closed lids as far as possible in the same direction as your legs are moving, unless otherwise indicated. At the end of each exercise, close your eyes suddenly and relax.

Relax in the Yoga Corpse Pose, as illustrated on pages 85–86, for at least a half-minute after completing each exercise.

Checkpoints: It is helpful to have someone observe your performance of corrective exercise. The following are points for you and/or your observer to note.

- When stretching the arms, are the elbows fully extended?
- When twisting the arms, are they fully twisted?
- When stretching the legs, are the knees fully extended?
- When stretching the Achilles' tendons, are they fully stretched?
- When performing movement on one side and then the other, is the movement fully and well executed on both sides? If not, on which side does the movement need to be repeated for the sake of improvement?
- In the repetitions, is the shorter arm and the shorter leg being consciously stretched?

Individual Corrective Exercises

1. Releases Tension in the Upper Body
Lie faceup with the legs open pelvis-width, and the hands behind the head with the fingertips on those (eye tsubo) points directly behind the eyes.

Inhaling, holding your power in the lower abdomen, stretch the Achilles' tendons. Exhaling, arch the chest up and push the elbows into the floor. Lift the heels an inch off the floor and hold the posture through a few complete breaths.

Repeat the exercise a few times.

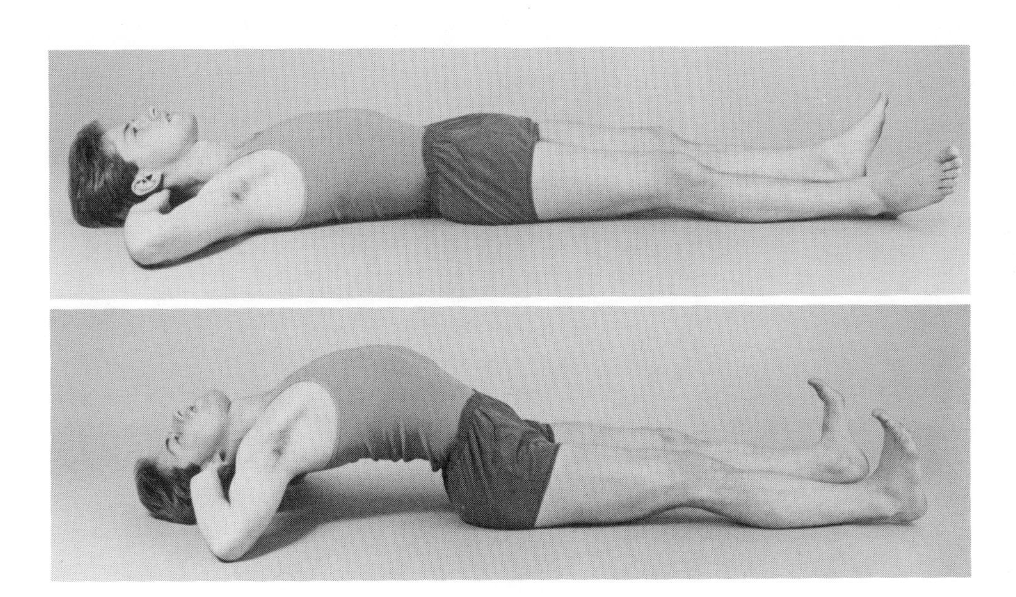

2. Relaxes the Chest
Lie faceup with the arms in an L-shape, palms up, with the fingers pointing away from the feet.

Inhaling, holding your power in the lower abdomen, stretch the Achilles' tendons and push the elbows into the floor. Exhaling, arch the chest up, bringing

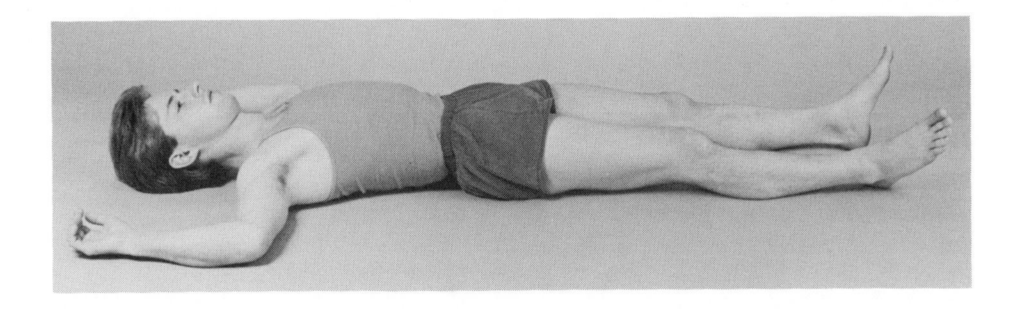

the shoulder blades together behind you. Inhale and lift the legs, looking upward. Exhale and lower the legs, looking downward.

Repeat the exercise a few times.

3. Releases Tension in the Neck and Shoulders

Lie faceup with the arms in an L-shape, palms down, with the fingers pointing to the feet.

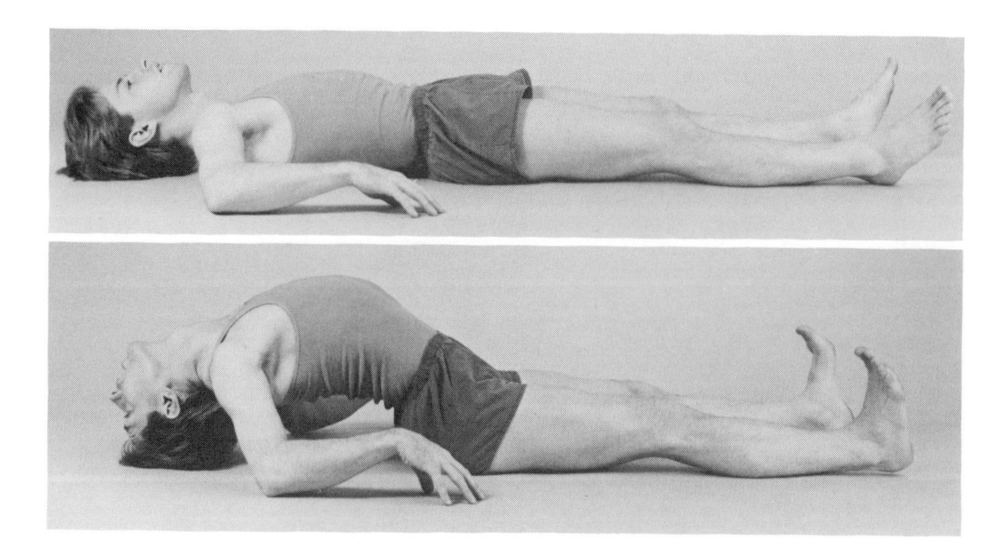

Inhaling, holding your power in the lower abdomen, stretch the Achilles' tendons and push the elbows into the floor. Exhaling, arch the chest up and lift the legs an inch off the floor. Roll your eyes upward and hold the posture through a few complete breaths.

Repeat the exercise a few times.

4. Increases Hara (Lower Abdomen) and Lumbar Power

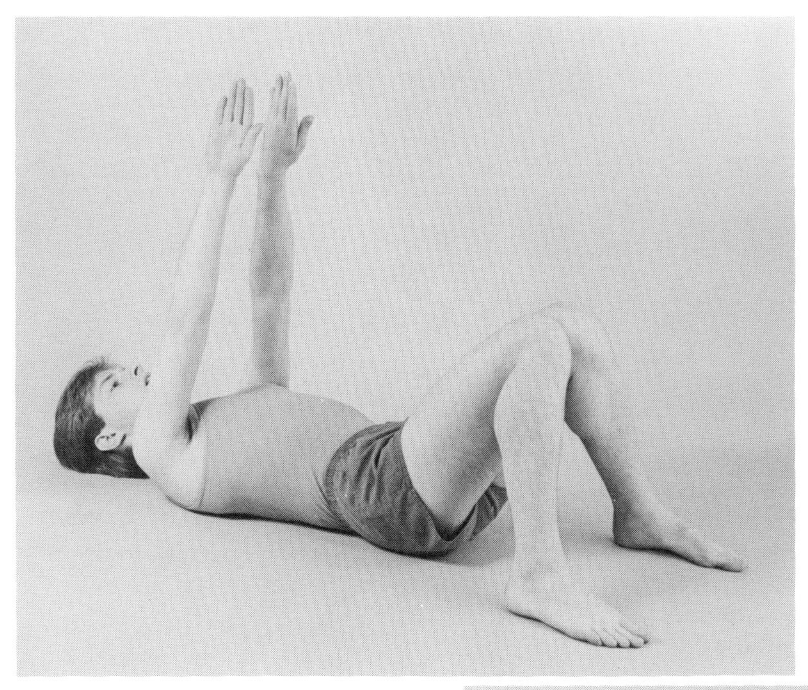

Lie faceup with the knees bent and touching, and the feet open pelvis-width. Stretch the arms straight up, keeping the backs of the hands shoulder-width apart.

Inhale and look at the ceiling. Exhaling, push the big toes into the floor, keeping the arms stretched and twisted inward, and slowly sit up as far as possible while rolling the eyes downward.

Repeat the exercise a few times.

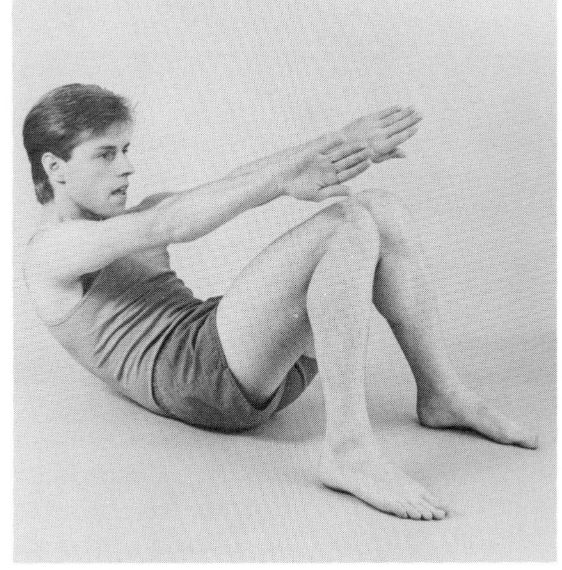

5. Strengthens the Hara

Lie faceup with the knees bent and touching. Place the thumbs on the first set of eye points, as depicted on page 42.

Inhale deeply. Exhaling, sit up while pressing your thumbs into the eye points. Continuing the exhalation, come forward as far as possible. Inhaling, come down slowly and relax.

Repeat the exercise five more times. For each repetition move the placement of the thumbs or index fingers to the next consecutive set of eye points, as depicted on page 43.

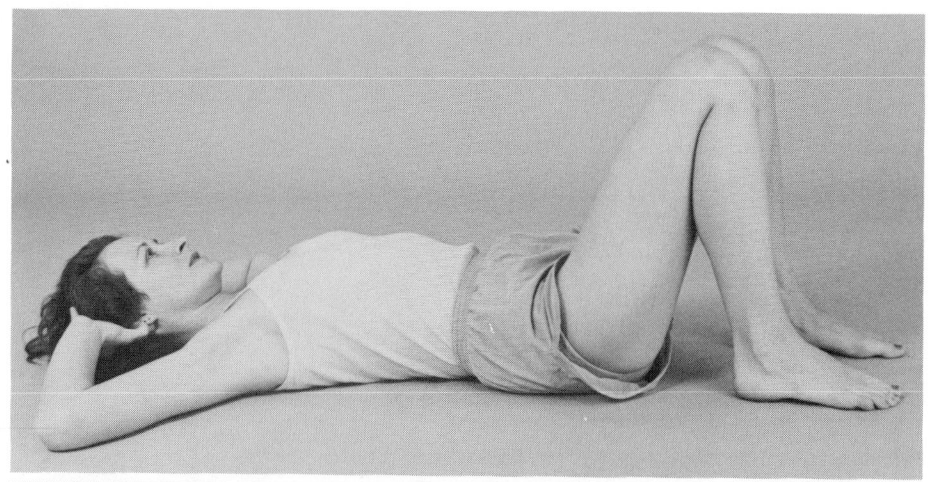

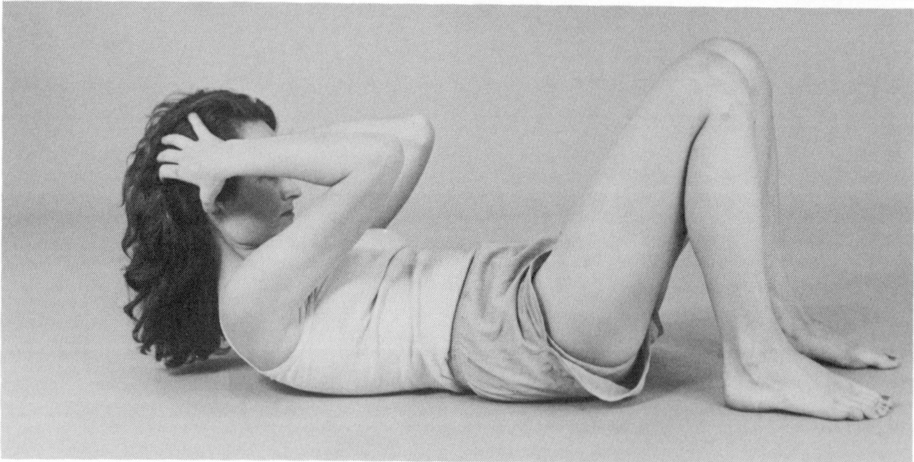

6. Relaxes the Entire Body

Lie faceup, bend the knees, and open them and the feet pelvis-width. Make fists with the thumbs tucked inside, and place the arms in an L-shape on the floor with the fists pointing away from the feet.

Inhaling, raise your hips. Exhaling, turn your head, letting your eyes follow

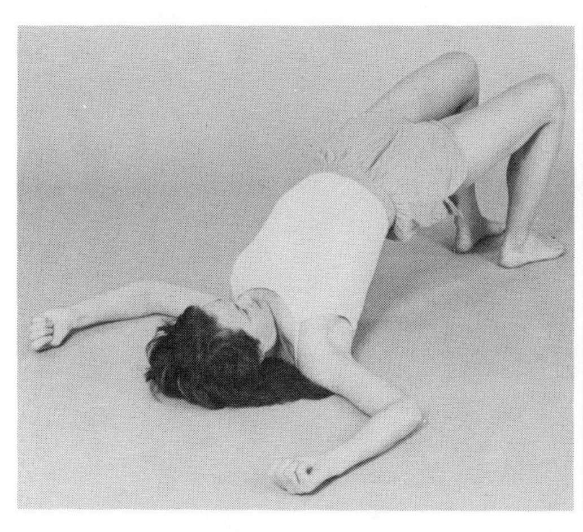

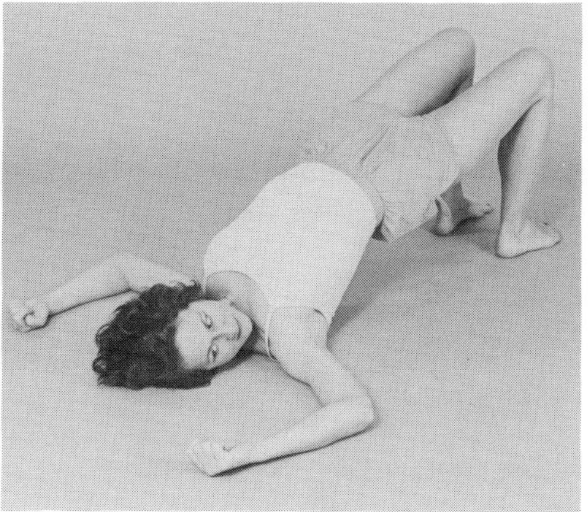

as far as possible to the right. Inhale and return to center. Exhaling, perform the movement on the opposite side.

Repeat the exercise a few times.

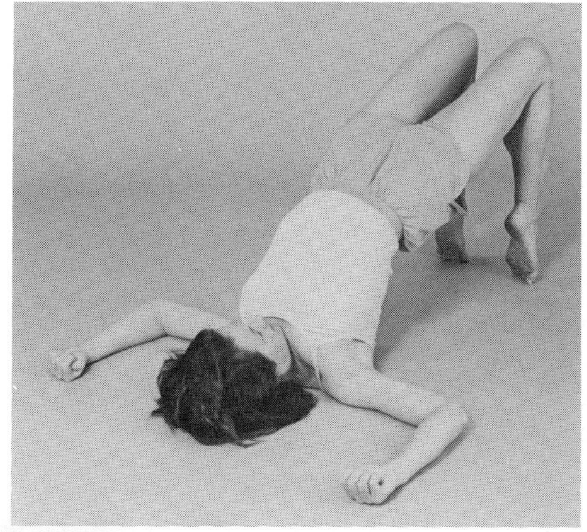

7. Relaxes the Entire Body
Lie faceup, bend the knees, and keep them and the feet pelvis-width. Make fists with the thumbs tucked inside, and place the arms in an L-shape on the floor with the fists pointing away from the feet.

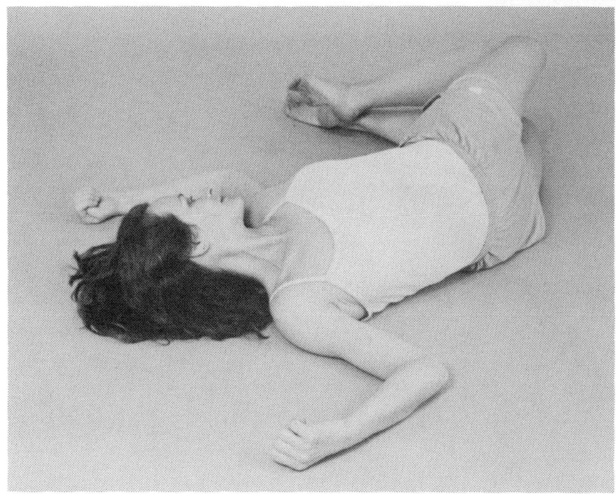

Inhaling, raise your heels and hips off the floor. Exhaling, turn your hips down to the right, while relaxing the neck and shoulders. Inhale and return to center. Exhaling, perform the movement on the opposite side.

Repeat the exercise a few times.

8. Opens the Chest and Relaxes the Upper Body
Lie facedown with the hands, palms down, beside the chest. Close your eyes.
Inhale. Exhaling, arch the torso and head upward and backward, while turning the hands outward and spreading the fingers. Open the chest, bringing the elbows together behind you, and look at the ceiling.
Repeat the exercise a few times.

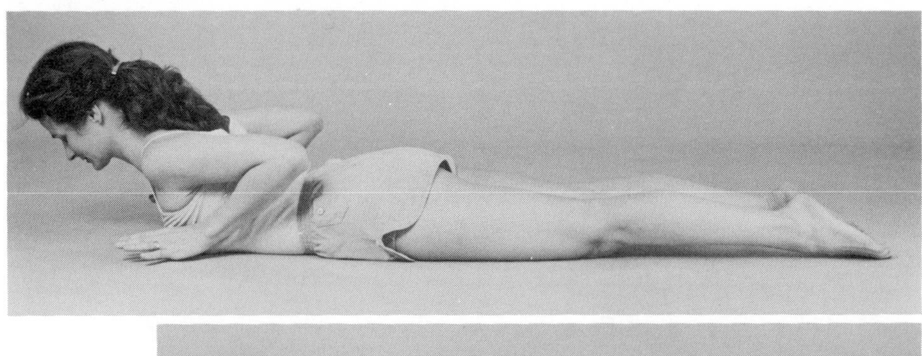

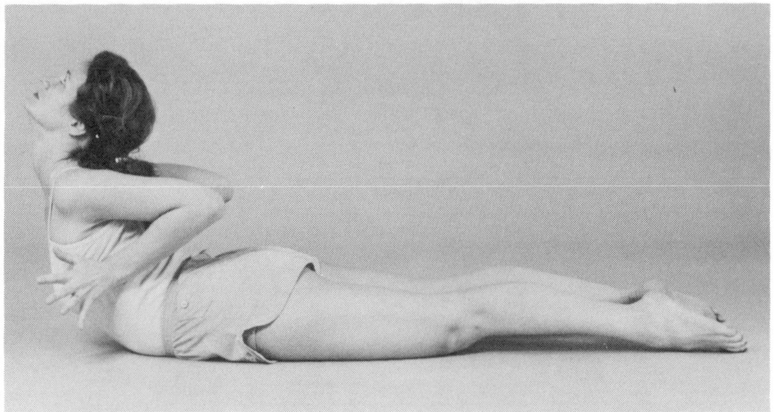

9. Strengthens the Lumbar Area and Relaxes the Chest and Upper Back.
Lie faceup with the chin on the floor. Bend the elbows and place your hands, palms up, near the waist.
Inhaling, slide the shoulders and elbows toward your head, bringing your hands

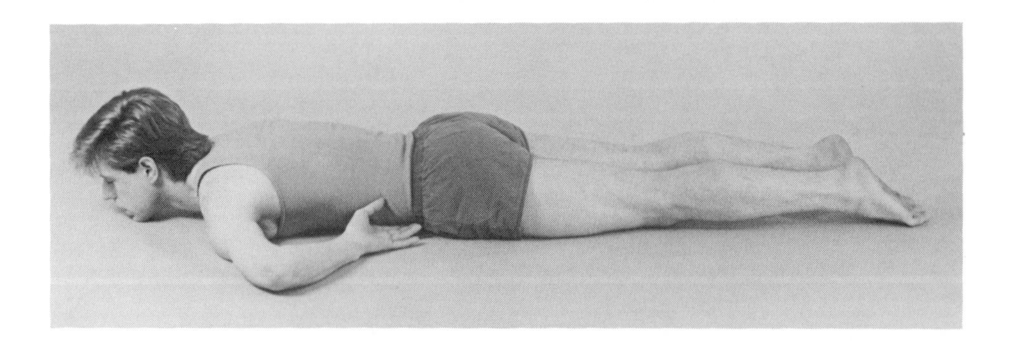

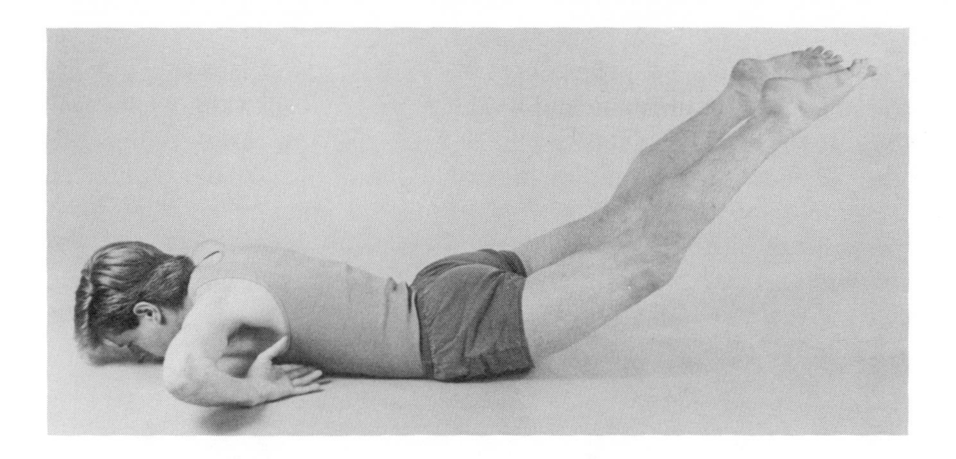

up to the armpits, while lifting the legs. Exhaling, slide the arms back to the original position, while lowering the legs slowly.

Repeat the exercise a few times.

10. Aligns the Spine and Pelvis, and Corrects Abdominal Imbalances

Lie face-down with the chin on the floor. Stretch the arms out to the sides, palms down.

Inhaling, holding your power in the lower abdomen, lift the right leg. Keep the shoulders on the floor. Exhaling, swing the right leg over the body and down near the left hand onto the floor. Stretch the right arm and relax the neck. Inhale and return to the original position. Perform the movement with the left leg and left arm stretched.

Repeat the entire exercise a few times.

11. Opens the Chest

Lie face down, bend the right knee, and reach back to clasp the right ankle with both hands, allowing the chin to come up.

Inhaling, arch the head and torso up, lifting the chin, and raise the right knee, pulling the right foot toward your head. Exhaling, turn the whole body down to the right and look to the right. Inhale and return to the original position. Exhaling, perform the turning movement down to the left, looking to the left.

Perform the movement with the left knee bent and left ankle clasped. Repeat the exercise a few times with each leg.

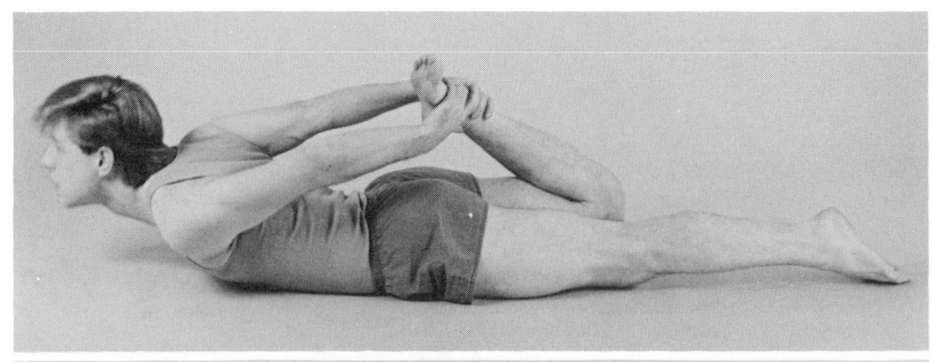

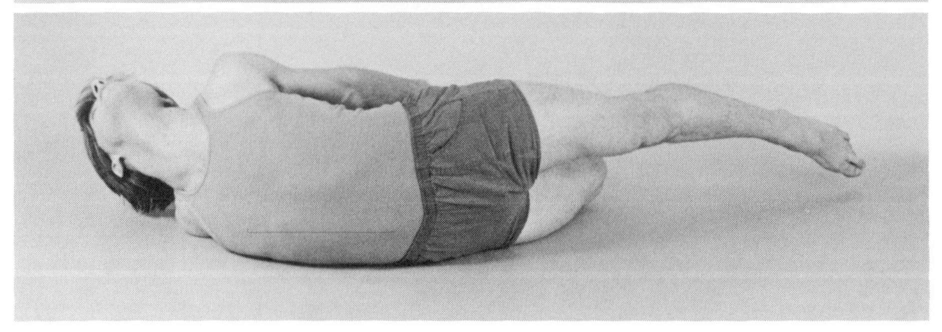

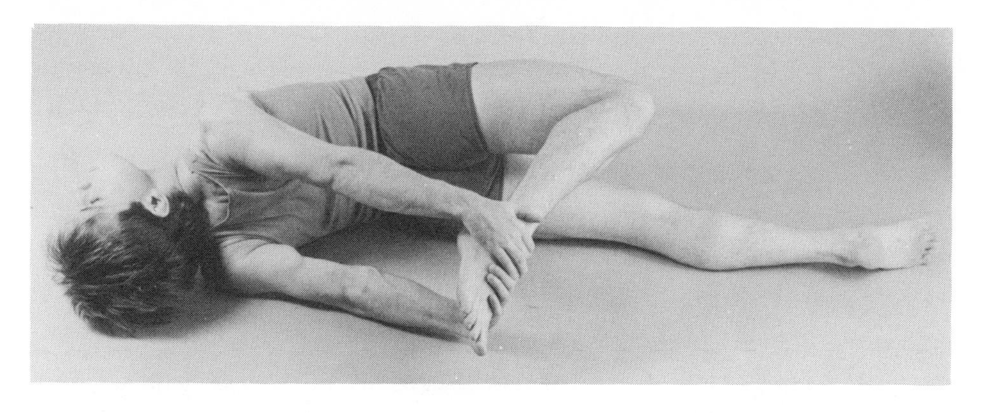

12. Balances the Lumbar/Pelvis Region

Assume the Yoga Plow Pose as illustrated on page 80.

Inhale deeply. Exhaling, keeping your toes on the floor, twist the hips to the right, while turning your head and eyes to the left. Inhale and return to center. Perform the movement to the opposite side.

Repeat the entire exercise a few times.

13. Strengthens the Lower Body and Opens the Chest
Lie faceup with the palms at the top of the rib cage, one on each side. Stretch the elbows out to the sides. Raise the legs straight up to ninety degrees, and stretch the Achilles' tendons.

Inhale deeply. Exhaling, keeping the shoulder blades on the floor throughout, lower the legs down to the right. At the same time, turn your head and eyes to the left. Inhale and return to center. Perform the movement to the opposite side.

Repeat the entire exercise a few times, twice as many times on the more difficult side.

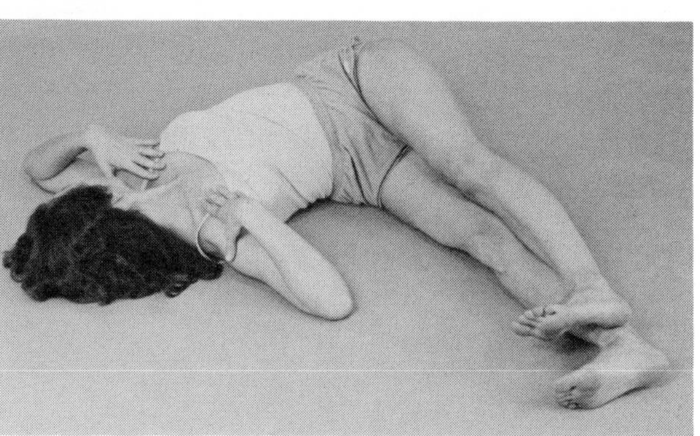

14. Strengthens the Legs and Releases Tension in the Neck
Assume the Yoga Cat Pose as illustrated on page 84 (A).

Inhale. Exhaling, lower the right shoulder to the floor, stretching the right arm out to the side (B). Inhale. Exhaling, stretch the left leg out past the right hand,

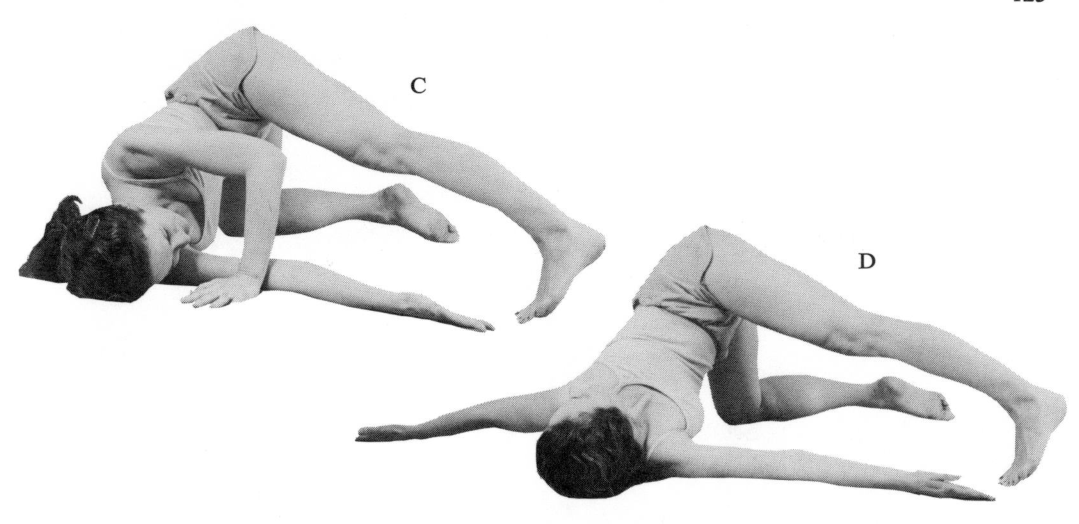

stretching the Achilles' tendon (C). Inhale deeply. Exhaling, turn the head and torso to the left, stretching the left arm out on the floor in line with the right arm (D). Look at your left hand. Hold the position through a few complete breaths.

Inhale and return to the Cat Pose. Perform the movement on the opposite side. Repeat the entire exercise a few times.

15. Stimulates the Lumbar Region and Stretches the Side and Thigh Muscles

Kneel upright with the knees open pelvis-width.

Inhaling, hold your power in the lower abdomen and clasp the right ankle with the right hand. Stretch the left arm straight up. Exhaling, twist the arm outward, push the hips forward, and arch the upper body backward. Return to the original position and perform the movement twisting the left arm inward.

Change the arm positions and perform the movement with the right arm stretched, outward and then inward. Repeat the entire exercise a few times.

16. Releases Tension in the Neck
Sit in the Seiza Posture as illustrated on page 134. Place the hands behind the neck, stretching the elbows open.

Inhale. Exhaling, drop the hips over to the left side. Inhale. Exhaling, twist the upper body, arms and head to the right. Look at the right elbow. Inhale and return to center. Exhaling, perform the movement twisiting to the left.

Perform the movement with the hips dropped to the right side, twisting to the left and right. Repeat the entire exercise a few times.

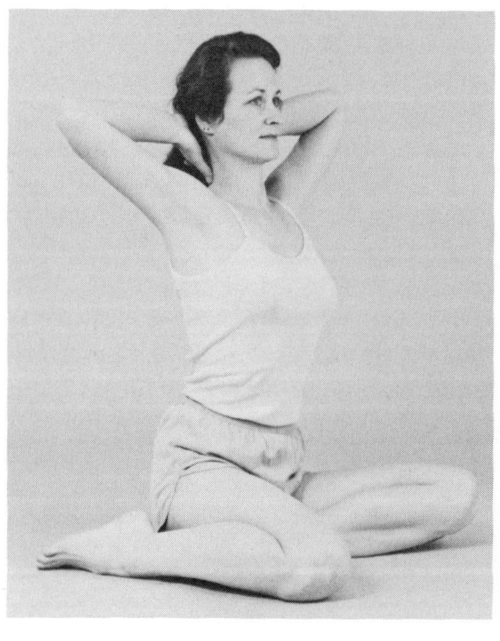

17. Opens the Chest and Releases Tension in the Neck
Sit in the Seiza Posture (as illustrated below and on page 134) or sit on the edge of a chair. Stretch the arms straight out to the sides at shoulder-height, and make fists with the thumbs tucked inside.

Inhale deeply, open your eyes fully, and arch the neck and upper body backward while twisting the arms outward. Exhaling, close your eyes, and bend the head and upper body forward while twisting the arms inward. Continuing the exhalation, bend all the way forward, touching the forehead to the floor.

Repeat the exercise a few times.

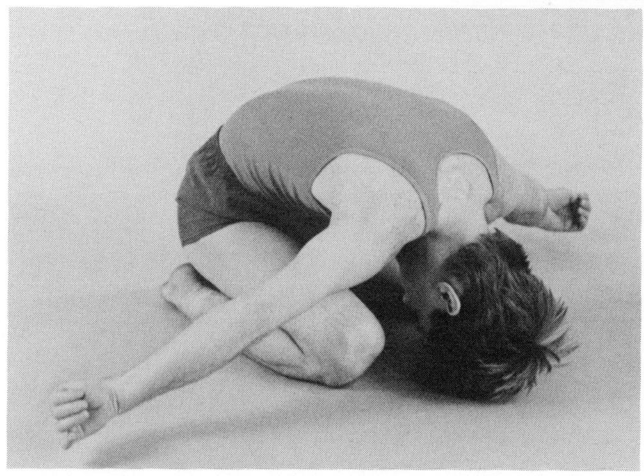

18. Stretches the Arms and Releases Tension in the Neck
Sit in the Seiza Posture (as illustrated below and on page 134) or sit on the edge of a chair. Stretch the arms straight out to the sides at shoulder-height, and make fists with the thumbs tucked inside.

Inhale deeply. Exhaling, twist the right arm outward and the left arm inward, while turning your head and eyes to the right. Look at the right hand. Inhale deeply and return your head to center. Exhaling, twist the right arm inward and the left arm outward, while turning your head and eyes to the left. Look at the left hand.

Repeat the exercise a few times.

Partners Corrective Exercises

1. Partner A lies faceup with the knees bent and the hands interlocked behind the head. Partner B holds the outside of A's knees.

Partner A inhales deeply, and, exhaling, sits up while opening his knees against an appropriate amount of resistance applied by Partner B. (An appropriate amount of resistance is such that requires effort but allows for completion of the movement.)

The exercise may be repeated several times, gradually increasing the amount of resistance applied.

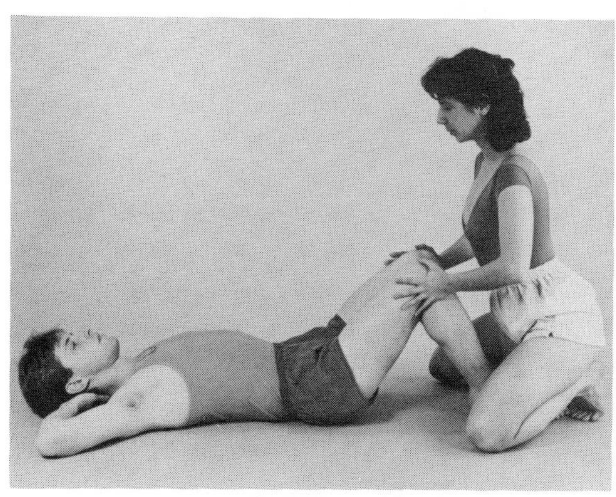

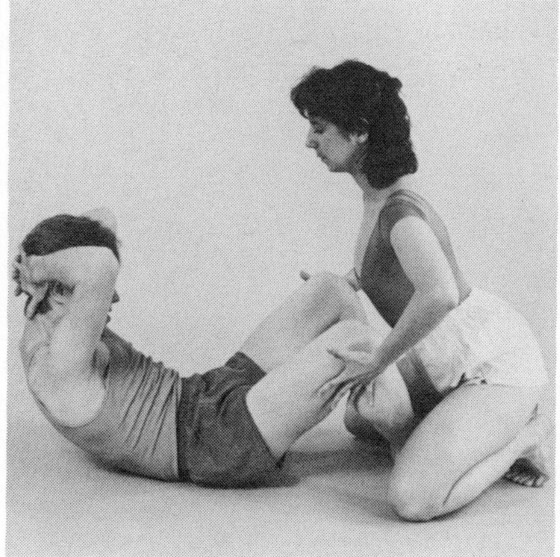

2. Partner A lies faceup with the arms in an L-shape, palms up. Partner B holds A's wrists.

Partner A inhales deeply, lifting his legs straight up into a right angle (to the floor), and stretches the Achilles' tendons. Exhaling, Partner A lowers his legs down to the right, keeping the shoulder blades on the floor, while turning the head and eyes to the left. Inhaling, Partner A lifts his legs to the original position and performs the movement to the opposite side.

The exercise may be repeated several times, alternating the leg movement from side to side.

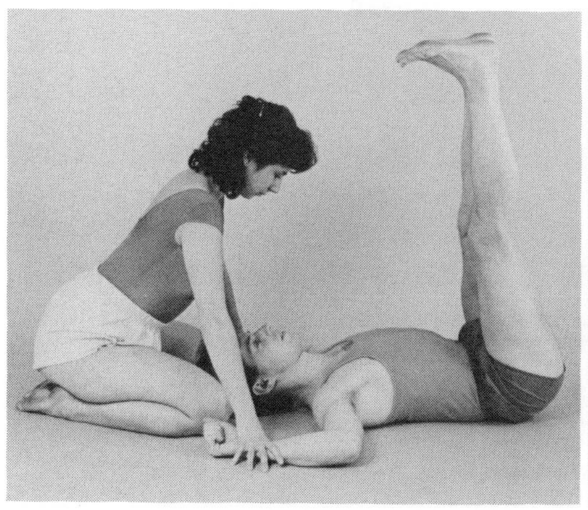

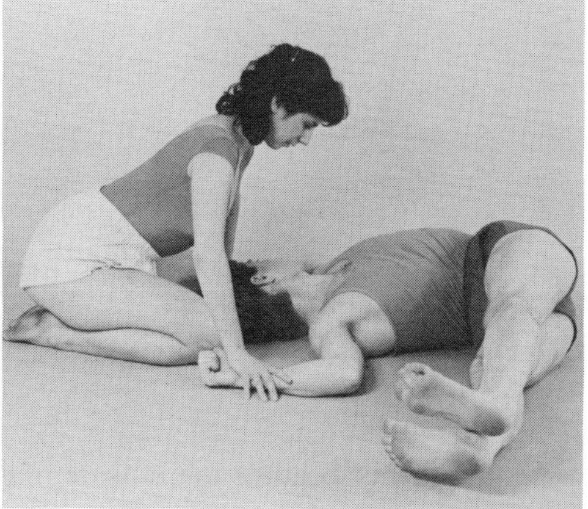

3. Partner A lies facedown with the arms and legs stretched on the floor in a line with the torso. Partner B sits lightly on the lumbar area of A, holding A's elbows from underneath.

Both partners inhale deeply. Exhaling together, Partner A arches his torso upward and backward, looking at his hands; while Partner B gently stretches A's arms backward and presses down firmly on A's lumbar area.

The exercise may be repeated several times.

Variation: Perform the same movement with Partner A's fingers interlocked. The variation may be repeated several times.

4. Partner A lies facedown with his hands, palms down, placed on the lumbar area. Partner B holds A's ankles.

Partner A inhales deeply. Exhaling, Partner A arches his torso up and back as far as possible, opening the chest by bringing his elbows together behind him. Partner A inhales deeply. Exhaling, Partner A twists the upper body and head to the right, looking at the right elbow. Inhaling, Partner A returns to center and performs the movement to the opposite side.

The exercise may be repeated several times.

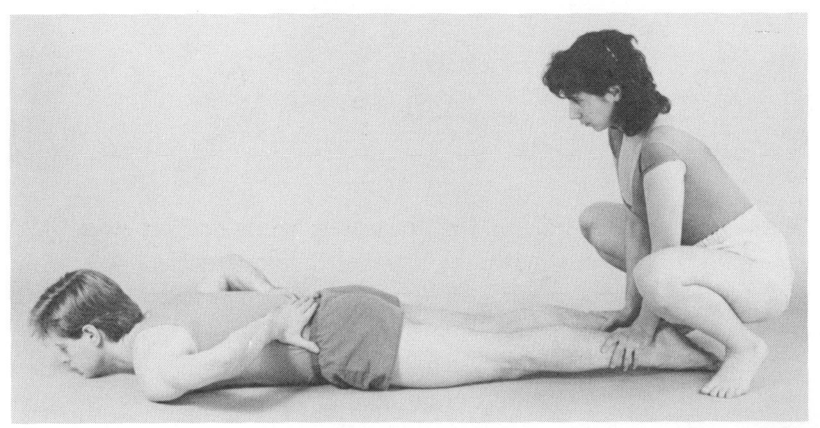

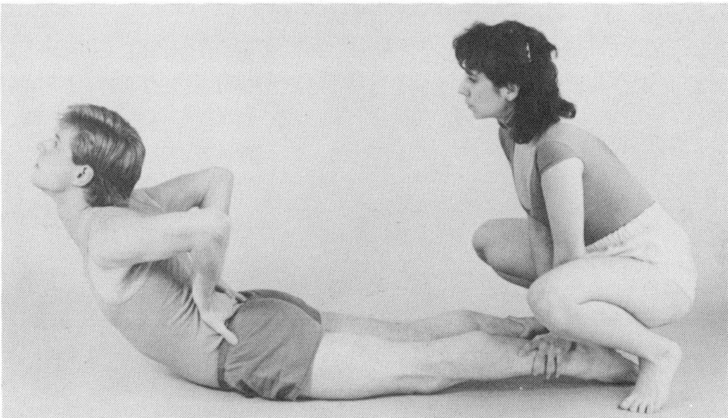

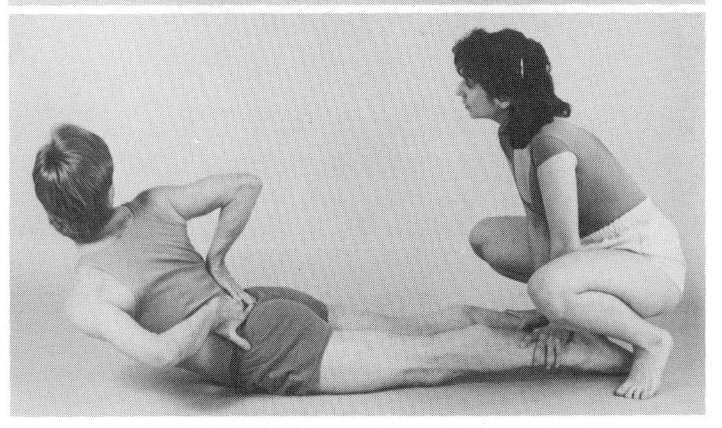

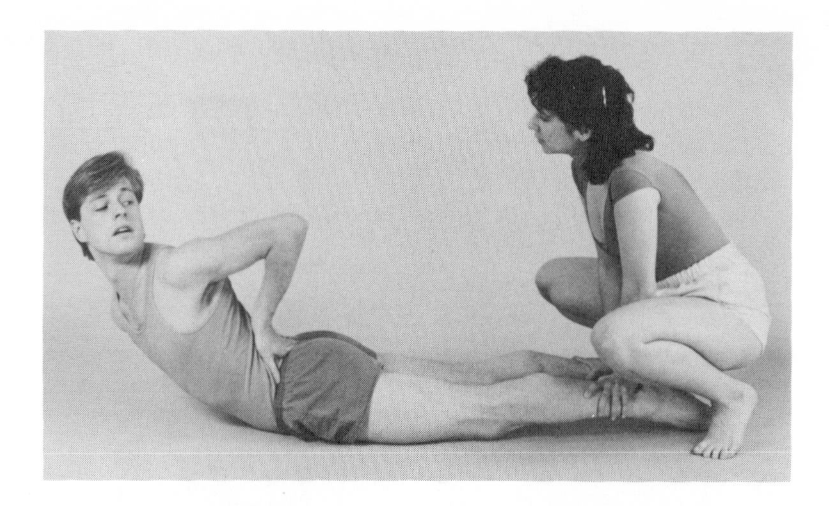

5. Partner A kneels in the Seiza Posture (as illustrated below and on page 134) with her hands, palms up, resting on the thighs. Partner B stands behind A, placing his left hand on the front of A's left shoulder and his right hand behind A's right shoulder.

Both partners inhale deeply together, and, exhaling together, Partner A twists her upper body to the right against an appropriate amount of resistance to the twist applied by Partner B. Just before the end of their mutual exhalation, both partners relax suddenly, so that A's body swings back to center automatically. The movement is then performed, with appropriate resistance, to the opposite side.

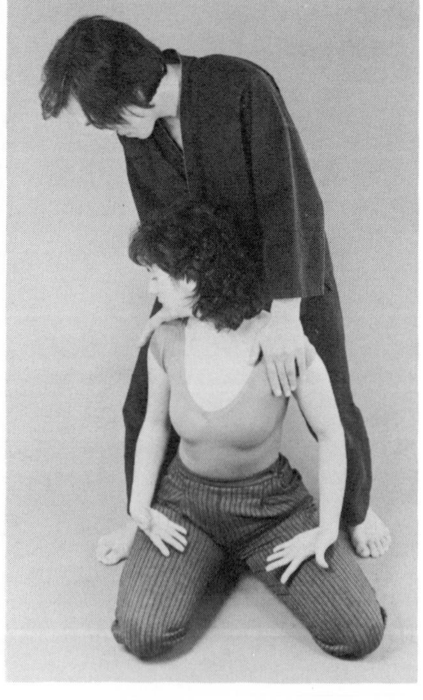

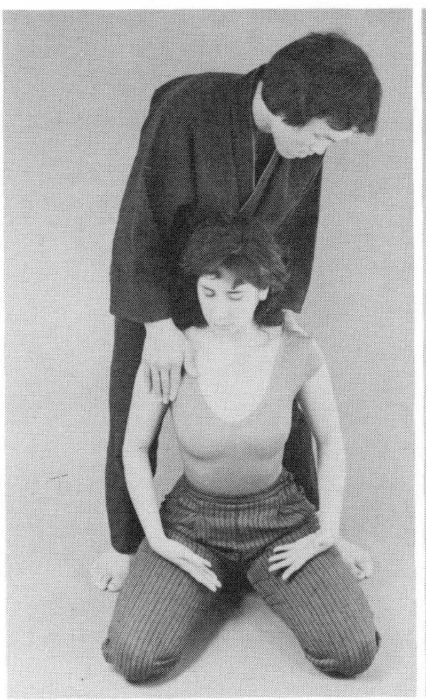

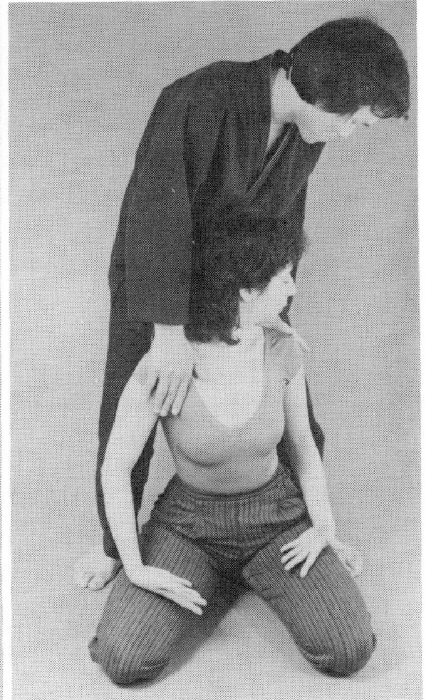

The exercise may be repeated several times, twice as many times to the more difficult side.

6. Partner A kneels in the Seiza Posture (as illustrated below and on page 134). Partner B stands behind A,

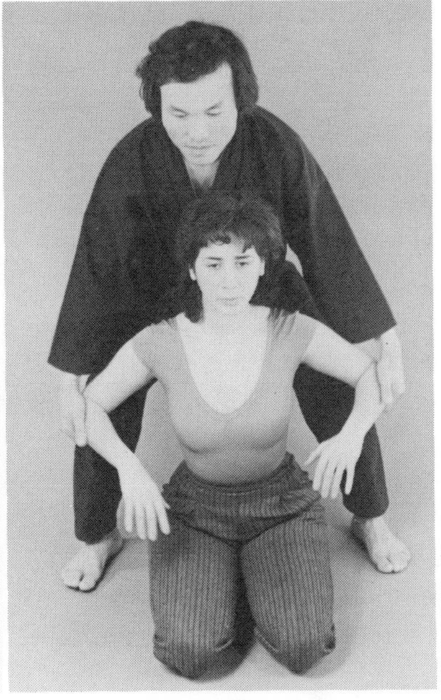

lifting A's arms into a right angle, her fingers pointing downward, and holding her elbows on the outside.

While both partners inhale together, Partner A raises her elbows as far as possible above shoulder-level. Exhaling strongly together, Partner A lowers her elbows against Partner B's resistance to her downward movement. Just before the end of their mutual exhalation, both partners relax suddenly, so that A's shoulders drop down automatically.

The exercise may be repeated several times.

Meditation

In recent times, meditation has become an integral part of the daily lives of millions of people. It is practiced for various reasons—purely as a spiritual discipline or for the expectation of a more personal result, such as mental relaxation. It seems unnecessary, then, to justify meditation. But it would be useful to examine an aspect of meditation that is not generally appreciated, which is its power to integrate inner vision with outer vision and to heighten both.

Just as there are different reasons for meditating, there are levels of meditation. Let us distinguish one level in particular, that which in Japanese is called *meiso.* The literal meaning of *meiso* is wide, especially in reference to one's way of thinking. *Meiso* as meditation is concerned with the greatest possible width or vision; that is, with seeing the visible and invisible worlds. *Meiso* also holds the implied meaning of to grasp what guides one at the moment in a specific place. In other words, the practice of *meiso,* in total, is seeing and combining the past, present, and future of the invisible and visible worlds in a moment.

Thus, the ultimate aim of *meiso,* the highest level of meditation, is to unify the mind, body, and spirit; which is to say, to reach the highest attainable state of human stability. Even though the source of this stability is within the invisible, inner center of the self, it can be reached—through experiencing, step-by-step, a process involving three stages of development. And these stages, presently to be defined, can be unified, or made one, by means of a three-fold form. In effect, the process of meditation involves combining content (the stages) with form.

Form and Content: Specifically, the three-fold form of meditation, the form that unifies the three stages of development, is: 1) deep breathing with a long, full exhalation; 2) natural posture with an upright spine and relaxed upper body; and 3) concentration with the mind focused though detached.

The practice of this three-fold form, which is the technical side of meditation, carries the practitioner through a realization of the aforementioned three stages of development, as follow:

1. *Dharana* (Yogic concentration): This involves the unification of the whole mind (conscious, subconscious and superconscious) with the whole body. It is reached by focusing all of oneself, mind and body, into what is called *hara* power,

whereby the lower abdominal and lumbar region is experienced as the center of being. *Dharana* becomes a part of daily life through the development of *hara* power, and is felt as a *yang* (active, creative) stimulation.

2. *Dhyana* (stability): This is a natural result of *dharana*. It involves physical relaxation and mental detachment, that is, detachment from emotion. *Dhyana* can be incorporated into daily life by approaching one's work as if it were a meditation. For example, even if you are busy on the outside, even if your surroundings are noisy, it is possible for your mind to remain calm and undisturbed. This is stability. On the other hand, if you are busy on the inside, even if your work place is tranquil, the surroundings will be experienced as agitated, noisy and oppressive. The state of *dhyana* in daily life, then, is felt as a *yin* (passive, receptive) stimulation.

3. *Hoge* (total release): This, the natural result of combining *dharana* with *dhyana*, involves unconditional giving. It is a state of complete openness and emptiness, a state of trustfulness without the need for protection; as such, it is the state of being that may be called prayer. When *hoge* is implemented in daily life, one is able to study without strain, to remember without trying, and to learn without effort. Essentially, *hoge* is a state of exultant *hara* power plus mental peace. It is felt as a balancing of active (yang) and receptive (yin) stimulations.

Unification or Hara Power: Since the above three stages commence with the development of *hara* power and culminate in the acceleration of *hara* power, it would be well to examine the nature of this phenomenon. *Hara* power may be thought of as physical and mental stamina. It is the foundation for the highest attainable level of human stability, and, as such, is the foundation of meditation. It resides in the lower abdominal and lumbar region; and when it is in force, the upper body, including the mind, face, neck, shoulders, and arms, is relaxed.

Hara power can be developed through attention to the following practices, all of which are part of the technical or structural side of meditation:

- closing the anus
- relaxing the muscles of the upper body
- stretching the spine upward
- breathing deeply and strongly from the lower abdomen
- focusing and calming the mind

When one's *hara* power is developed, the midbrain becomes balanced. Since the midbrain controls the functioning of the autonomic nervous system, then *hara* power affects the workability of the internal organs and the modulation of instinct, emotion and desire.

Thus, if you pay attention to the technical points of meditation for a period of time, that is, if you persist in improving the three-fold form of meditation, your *hara* power will be increased. Concurrently, you will undoubtedly reduce your food intake, become in command of sexual desire, and gain control of your emotions.

Hara power, established through practicing the structural aspects of meditation, is the essential ingredient of meditation. Through its development, you can combine concentration (*dharana*) with stability (*dhyana*) culminating in total release (*hoge*). And when these states of being filter through into daily life, you are living in what is called natural or Buddha mind.

The Postures of Meditation

Successful meditation is basically combining correct breathing with correct posture. When the breathing is correct, the mind will be focused; and when the posture is correct, the *hara* will be strengthened. Before meditation, practice some deep breathing exercise. Then, composing yourself in one of the postures for meditation, imagine that a string, rooted in the bottom-most vertebra, is pulling your entire spine up into the sky. At the same time, pull your pelvis backward. Allow your shoulders to drop down and relax. Pull the chin in, stretching the back of the neck, and look down at an approximately forty-five degree angle with your eyes half open. While breathing deeply, making the exhalation longer than the inhalation, center your strength in the lower abdomen (*hara*) and keep the upper body relaxed. Hold your selected meditation posture, chosen from among the following, for at least fifteen to twenty minutes.

Seiza: Kneel with the hips on the legs, the knees together, and the big toes touching each other behind you. Stretch the arms straight out in front of you, and place the palms on the floor. Push your hips straight back, stretching the spine. Inhaling, raise the upper body, bringing the spine upright. Raise the knees, allowing them to fall open naturally to approximately one and a half or two fist-widths. Keep this distance between the knees and lower them slowly. Place the hands, palms up, on the upper thighs and join the thumb with the finger of each hand, making a circle. Keep the elbows out from the body. Stretch the spine up from the coccyx. Tuck the chin in, stretching the back of the neck, and relax the shoulders.

Butterfly: Sit on a small cushion or pillow. Bend the knees, bringing the right heel into the crotch and the left heel into the right ankle. Keep both knees on the floor. Place your hands on the thighs, palms up. Stretch the spine and neck upward, open the chest, and relax the upper body.

Half-Lotus: Sit on a small cushion or pillow. Bend the knees, bringing the right heel into the crotch and the left foot onto the right thigh. Keep both knees on the floor. Place your hands as in the quarter-lotus posture.

Lotus: Sit on a small cushion or pillow. Bend the knees, tucking the right foot into the left thigh joint and, placing it on top, the left foot into the right thigh joint. The cushion under you must be high enough to enable both knees to touch the floor. Place your hands as in the quarter-lotus posture.

Sitting: Sit on the edge of a chair with the knees opened pelvis-width, the lower legs straight, and the feet parallel, flat on the floor. Maintain your power within the big toes and the insides of the knees. Stretch the spine and neck upward, open the chest, and relax the upper body. Place your hands as in the quarter-lotus posture.

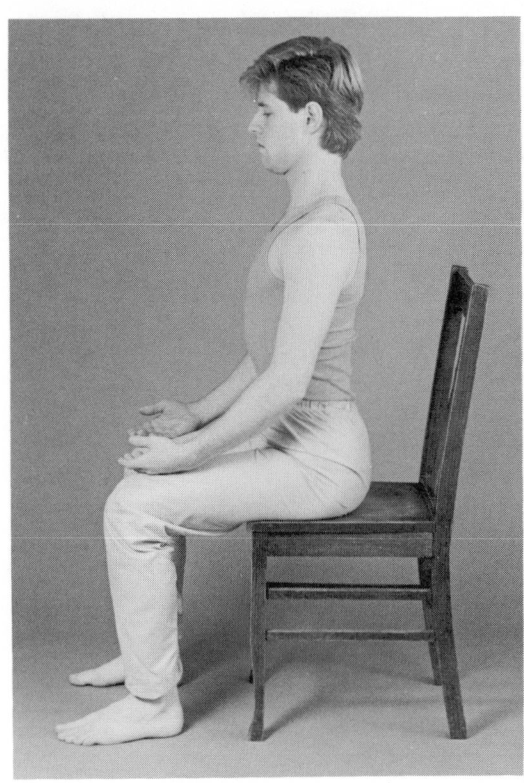

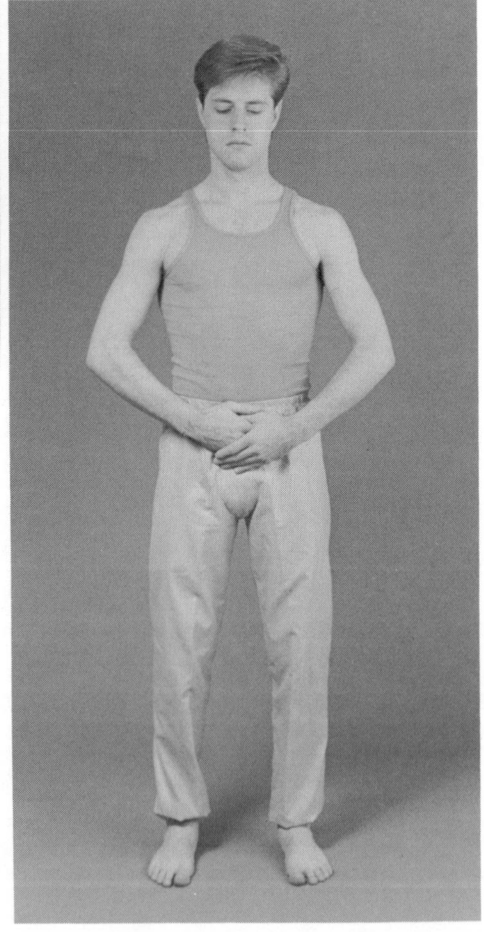

Standing: Stand with the legs open pelvis-width and the feet parallel. Tuck the chin in, so that the nose and navel are in one line and the ears are over the shoulders. Push the big toes into the floor. Place the edges of the hands, one palm on top of the other, at the pit of the abdomen just below the navel. Interlock the thumbs. Extend the elbows out to the sides, allowing the center of gravity to move down into the *hara*.

Meditation and the Eyes

As you meditate, you will be deepening your breathing, taking in a greater supply of oxygen than you are probably accustomed to. Since the eyes are second only to the brain in the amount of oxygen required, meditation is a good way of feeding your eyes a fuller supply of oxygen. Also, as you meditate, you will be increasing your *hara* power—through which the functioning of the autonomic nervous system, at one end of which are the eyes, will be improved. Furthermore, as you meditate, you will be enabling your upper body to relax—through which the entire nervous system, including those parts innervating the eyes, will be revivified, tending toward freedom from paralysis.

To relax the eyes even more directly through meditation, you might wish to practice one of the meditations for vision improvement. To do so, take a lighted candle or a sheet of white paper marked with a bold black dot. Place this object before you, below your direct line of vision, so that the eyes must look slightly downward. Assume one of the postures for meditation, and concentrate on the flame or on the black dot while breathing calmly. When tears come to your eyes, or when your eyes become tired, close them slowly. When your eyes feel rested, open them, focusing on the object once again, and continue the meditation as above for as long as you wish.

Detachment and the Eyes

The stage of development known as *dhyana* or stability, in which the body is relaxed and the mind is detached from emotionality, is a state of existence that includes within it how one uses his eyes. To see clearly what is there for you to see, that is, to maintain a natural relationship with the external world, is to combine inner vision with outer vision. It is to see with detachment. If you can train yourself to see without struggling to see, you will avoid eyestrain and your power of vision will become stable.

The natural, the true and the ultimate way of healing is through detachment. To heal your eyesight, you must let go of those enslaving conditions or habits that you have become dependent upon, that have caused your eyes to suffer and your eyesight to weaken. When these conditions and habits are released, when you can detach yourself from them, when, finally, you can change your life-style, your healing power will come alive and your eyes and vision will change.

This philosophy of non-attachment is the fundamental philosophy of Yoga. In practice, it means being open always to new experience, to new stimulation, to change—in order to discover what you need for enlightenment.

Chapter 6

Questions and Answers

All the following questions were proposed by people attending the Oki Yoga Dojo in Japan for the purpose of studying or to heal their vision. The answers, formulated by Masahiro Oki, are here to serve as a summary encapsulating the major ideological and practical principles of this book.

Q. What is the meaning of nearsightedness and eye disease?
A. Nearsightedness and degenerative eye diseases take a long time to develop. They are chronic diseases. Any chronic disease is not simply an illness, but is representative of sickness in the life-style and character. To improve the eyesight, you must change your character—body and mentality—through changing your life-style.

Modern medicine does not approach vision problems or any other chronic disease in this manner. Nevertheless, the only way to improve chronic disease permanently is to change the life-style.

Q. What is the strong point Yoga has to offer toward improving the eyesight?
A. Yoga is based on *mikkyo*, which is an attitude of mind that does not criticize or deny the other. It also does not embrace any particular way or method. Yoga can therefore accept, apply and combine the positive features of all variety of ways and means, teaching the utilization of that which is appropriate and necessary.

From living through a variety of life experiences, I am aware of myriad healing techniques and theories. Each method makes some sense. From each type of therapy, some people derive benefit. But if you depend upon one type of therapy only, say, diet or exercise or acupuncture only, or if the measures you take are but partial—say, doing eye exercises only and ignoring your breathing, the process of healing will be necessarily handicapped and your mind will continue to be dependent upon fixed measures.

Yoga teaches impartiality and independence. Even though you might expect to find a magic formula for perfect eyesight, in fact, the methods and theories presented are not directed toward improving the eyesight only. They are directed toward an improved and integrated use of the body and mind. Each individual is encouraged to extract those parts (of the book) appropriate and necessary for his body and mind, for his vision improvement. In this sense of Yoga's being aware of individual necessity and appropriateness, Yoga can be said neither to condemn nor condone anything.

Q. What do you think about using modern medicine and/or traditional Oriental medicine to improve the eyes?

A. Yoga combines philosophy and discipline. For the individual, Yoga consists of a search for what is needed to reach *satori* (enlightenment).

Modern medicine and Oriental medicine each have strong points and weak points. Absolute good is non-existent. You should take from both kinds of medicine that which best aids your vision. For example, certain herbs help the muscles of the eyes relax. Shiatsu and acupuncture can bring on vision improvement. Modern medicine is expert in conducting scientific examinations and identifying disease. It then recommends medication or surgery or some application, like the wearing of glasses, to alleviate the symptoms of the ailment or disease. But if you depend upon modern medicine or Oriental medicine alone, or even in combination, you will never cure any chronic disease completely.

To correct your vision, you must first heighten your ability to heal yourself. You must practice disciplines that are holistic, that integrate the body and mind. These disciplines can of course be complemented by various methods of Oriental and/or modern medicine. For example, you might use modern medicine to have your eyes examined, and use Oriental medicine to help relax your eyes and improve blood circulation to them. But all the while, you should use the philosophy and practice of Yoga to improve your power to heal yourself.

You can certainly wear glasses at any time necessary to adjust your vision momentarily, particularly if your eyesight temporarily worsens during the process of developing your power to heal yourself. But in general, modern medicine's viewpoint on nearsightedness is to wear glasses or contact lenses all the waking hours (and now, in the case of some contact lenses, even all the sleeping hours), though neither of course improves the eyesight. The viewpoint of Yoga is that nearsightedness is an opportunity to develop your power to heal yourself. It is an opportunity to change yourself.

Q. Could you explain the meaning of pseudomyopia?

A. Pseudomyopia, which is not true nearsightedness, is due to temporary circulatory trouble. If the eyes only are examined, the condition appears identical to actual myopia. But more than the eyes must be examined to determine if the condition is actual or pseudo.

Any imbalance in the blood circulation means that there is also muscular imbalance. Muscular imbalance is misplaced tension. Most frequently, the primary cause of pseudomyopia is poor posture. For example, if you hunch over while reading, the upper body becomes tight (misplaced tension) and the eyes congested (circulatory trouble). The first step in curing pseudomyopia is to consider how to relax the upper body and improve the blood circulation overall. The various postures assumed for daily life, the use of the eyes, and the mental attitude must be checked and prospectively changed. The diet too needs consideration. If you eat refined sugars or too many sweets, not only the teeth but the eyes also are harmed.

Yet pseudomyopic people, not understanding their condition, resort to wearing glasses. Wearing glasses then becomes the main reason why their vision worsens.

Q. What kind of diet is good for the eyes?

A. There is no food that is good for the eyes alone. If a particular diet is good for the whole body, it is also good for the eyes. Fundamentally, nutrition should be balanced. A balanced diet is one that allows nutrients to be absorbed, and digested food to be eliminated. It allows the body to cleanse itself, and allows the blood to maintain the correct acid/alkaline balance.

The eyes have many nerves surrounding them. These nerves need vitamins, minerals and oxygen. From the nervous system's point of view, vegetables are preferable to animal food. Also, if you eat primarily whole grains and vegetables, the blood will never become too thick, the blood pressure will never rise too high, and you will never tire too easily.

Specifically, vitamin B is beneficial to the eyes. Brown rice contains plenty of B vitamins and therefore is helpful to the vision. Vitamins A and C also help the vision, though an excess of vitamin A can contribute to the development of cataract.

On the whole, you must take food which has the qualities you need. You must not deduce that since vegetables are good then meat has to be bad. It is a simplistic way of thinking. In India, the Mohammedan people eat meat while the Hindus are vegetarians, but their compared percentages of nearsightedness and eye disease are not significantly different.

The main point is that the blood must not be overly acidic—from unbalanced nutrition, from poor posture, from inadequate movement, or from negative mental attitude. The eyeballs of overly acidic people are too soft, which leads to the wearing of glasses. Find for yourself the food and training that maintains the correct acid/alkaline balance in your blood.

Q. There seems to be a great increase of nearsightedness among children nowadays. Why is that?

A. In some children, nearsightedness is a symptom of senility. If a child's life is overly protected, if he is allowed little exercise, given refined foods and sweets, and if his surroundings contain mental anxiety, he becomes weak. A child denied adequate playtime and play space and pushed to study becomes irritated. His life-style is, in effect, similar to an adult's and is unsuitable for a child. A child with an adult life-style manifests geriatric disease such as weak eyesight.

Also, people living in cities are usually far removed from Nature. Without the influence of Nature, a child's body and mind, because they are growing and developing and are not yet strong, can easily become unbalanced. On the other hand, children can easily recover from imbalances, including nearsightedness. It has been surprising to see how many first and second grade children readily cure their eyesight at the Oki Yoga Dojo.

Another factor is that when a child complains he cannot see well, his parents take him to a doctor for eyeglasses. Yet at this stage the child's problem is not actually poor eyesight, but poor blood circulation due to life-style, including diet and posture. This condition of pseudomyopia, which could easily be changed without the use of glasses, is confused with myopia.

Q. Is nearsightedness hereditary?
A. You are of course descended from your parents, and so have some similarities in body structure, constitution and character. Yet it is very difficult to know for sure just which of your qualities are based entirely on heredity.

It might seem to you that nearsightedness and vision problems are hereditary. But if a quality is hereditary, except for color blindness, it most certainly cannot be changed. Yet, in most cases, vision problems can be changed and cured completely. There might be a hereditary tendency for eye weakness in a person, but nearsightedness itself, or any eye disease, is rarely, rarely hereditary.

If you change your life-style, especially change it from that of your parents, you can then discover which of your qualities are and which are not hereditary. If your parents are nearsighted and you are too, it might be that you have imitated major aspects of their life-style, and nearsightedness is one result.

Q. Could you explain astigmatism, particularly in terms of a right side and left side imbalance?
A. In astigmatism, the stimulation being received by the right side of the body is different from that being received by the left. This leads to the imbalance in the vision. Also, the shoulder heights of an astigmatic person are different, as are the arm strengths. Imbalance between the right and left sides regarding the vision is a whole-body imbalance. To improve astigmatism naturally, you must correct the balance of the whole body.

Q. Why are nearsighted people narrow-minded in general?
A. Nearsighted people alone are not narrow-minded. Any kind of chronic disease can lead to narrow-mindedness. A narrow-minded person is one who easily becomes tense, so it cannot be said that only the minds of nearsighted people are narrow.

All disease stems from an imbalance in the nervous system, producing tension, and an acid/alkaline imbalance in the blood, producing a lack of energy. The main benefit of natural healing is that the person is able to relax. But relaxation must not be confused with powerlessness. To be relaxed is not to be slack or languid. True relaxation is to be calm, peaceful, not loose, and to have energy.

A life-style based on Yoga improves the functioning of the parasympathetic nervous system, allowing the muscles to be freed of tension, and increases the alkalinity of the blood, allowing more energy. That is why the whole basic theory regarding vision correction is to combine exercise with fasting. Fasting helps the muscles relax and purifies the blood. The organs can then begin again to function well and the nervous system to wake up. Exercise assists and maintains the renewed condition.

Q. Is it better to wear glasses or contact lenses or none at all?

A. Glasses have their own characteristic points. Contact lenses have theirs. In general, it is unnecessary to wear glasses and/or contact lenses continuously. If you do wear them, further weakness will arise.

A good use of glasses is to wear them when necessary and remove them when not. Wearing glasses or contact lenses all the time disallows the eyes a chance to relax. Wearing glasses is to the eyes as carrying luggage is to the arms. When you put down the luggage, your arms can relax. When you take off the glasses, your eyes can relax.

It might be that when you remove the glasses you experience a tension headache, created by trying to see what you cannot. This is a sign that the body is attempting to adapt to life without glasses, but that the sudden removal of the glasses might be making adaptation difficult. It is the same in the case of food. If a person regularly overeats and then suddenly begins to fast, he will experience extreme reactions. If when removing your glasses you experience pain or must make an extreme effort to see, then do not dispense with glasses suddenly and all at once. Rather, remove them for a period, put them back on, remove them again—extending the removal time all very gradually.

As your vision improves, you should have the prescription of your glasses changed. If you cannot afford to do this, then try wearing the glasses for as short a time as possible, only when absolutely necessary. I am not saying that you should not wear glasses. I am saying, according to the viewpoint of Yoga, that when you need to visit an eye doctor, you go; and when you have an organ imbalance, you heal it. You wear glasses when you need them, and you train yourself to heal your eyesight. Yoga does not recommend an easy or partial way. It teaches to do as is needed, step-by-step.

Q. Do people who have cataracts or glaucoma easily become nearsighted, and what do you think of cataract surgery?

A. It takes a long time to develop cataracts or glaucoma. Before that, the eyes are undoubtedly weak already. It is quite possible that nearsightedness is present or soon will be. In actuality, cataracts and glaucoma are conditionally the same as myopia, which simply means there are three types of nearsightedness—the cataract, the glaucoma, and the weak-eye type. In all three, due to the life-style, the eyes are dull and interocular pressure is high.

The problem with cataract surgery is that the post-operative condition of improvement might not last. Cataract surgery, by removing the lens, eliminates the cloudiness of the eye. But it does not correct the cause of the cataract. The main point to understand is that the cause of eye disease is not only the eyes. An eye doctor cannot cure nearsightedness or alleviate the cause of cataract. An eye doctor can prescribe glasses or perform surgery, but he is not healing the vision.

Q. Could you explain why some people with kidney or liver problems are nearsighted and some are not?

A. There are various classic body types. In each type, some parts of the body are strong and other parts are less strong.

People whose eyes tire easily usually have weakness in all five senses. They are prone to nose and ear, as well as eye problems. If you practice corrective exercise for nearsightedness, you will see that your nasal passages and ears feel better too.

Most eye problems are related to some internal organ disturbance. If the liver or kidneys are weak, the blood quality becomes unbalanced. If the kidneys are malfunctioning, the blood pressure within the eyes becomes unbalanced and the strength of the bones, neck and brain is minimized. If the stomach or liver is malfunctioning, the neck and/or shoulder blades become unbalanced. All these imbalances result in poor stimulation to the eyes. In particular, when the shoulders are burdened, due to constipation or overeating, for example, the eyesight diminishes. You might have noticed that when you are hungry, your shoulders become light and your vision improves.

Q. When I fasted, my eyesight became worse. What is the reason?

A. During fasting, blood collects in the weakest or most ailing part of the body. This is the body's way of trying to strengthen or heal that part. If you have some trouble more severe than nearsightedness, perhaps in one or more of the internal organs, the blood will concentrate itself in that part during fasting. There is then a shortage of blood supply to the eyes. This phenomenon of hazy vision during fasting often happens to people with cataract or glaucoma.

Q. Why is it beneficial to the eyes to twist the arms as a part of corrective exercise?

A. When you twist the arms, the muscles stretch. Stretching the muscles releases tension from them and improves the blood circulation. Perhaps you have noticed that when the body is twisted, it becomes warmer. Also, when you twist the arms, the joints contract. Within each joint there are muscles and several tsubo points. So by twisting, these tsubo points are stimulated, awakening the nervous system. To combine twisting with stretching is more effective than to stretch only.

Q. When a person is happy or satisfied, why do his eyes shine?

A. When you are happy or feel good, your autonomic nervous system becomes balanced and the hormone system can function well. That is why, when you are happy, the whole body feels refreshed, having a tingling sensation of moisture. It is through this sense of moisture that the eyes shine.

The eyes reflect the state of mind. If you observe a person's eyes, you can read his mental condition. The eyes of a neurotic person, for example, are like the eyes of a dead, decaying fish. This is because his autonomic nervous system and hormone secretion are unbalanced.

The reason Yoga makes mental relaxation the first step in healing the vision is because if your mind is accepting and joyful, you do not become so easily unbalanced or fatigued. When you can learn to use your eyes with enjoyment and to relax your mind, the whole body and mind will become balanced.

Q. What kind of mental attitude should we have toward life?

A. None of us made a conscious decision to be born. Nature brings out what it needs and discards what is unnecessary. Universal law encompasses everyone and everything. To create anyone or anything on this earth, Nature causes the right components to come together at the right time. This phenomenon may be called destiny. And it is destiny that makes it irrelevant for anyone of us to try to think of a reason why we were born. The fact that we were born is sufficient reason for being born.

Nature offers us varied opportunities and alternatives throughout life. We have the opportunity to grow at all times. We have the possibility to be happy, to be sad, to meet, to separate, to profit, to lose. As these potentialities come to us, and as we are capable of understanding that these varied stimulations are especially for us, we are able to see in every experience a lesson and an opportunity to grow. It is possible to grow every second of one's life, to grow from each different stimulation that comes.

Sometimes an outside stimulus will change your life in a second. Sometimes change takes place slowly because you are insufficiently developed to accept a lesson needing to be learned. What is fundamentally important is to train yourself to accept every stimulation, every opportunity, with gratitude. By putting maximum effort into using each moment with which you are entrusted, you affirm your faith.

When we get sick, it is because we have to get sick. Sickness is not unreasonable. Whatever stimulation we are offered, whether we like it or not, is an opportunity to be taken advantage of completely, an opportunity to give of ourselves to the utmost. This is the way to live.

Appendix

The Tibetan Eye Chart was designed by Tibetan Monks to exercise the muscles of the eyes. The Tibetans have used natural eyesight improvement exercises for centuries. If the chart is used a few minutes twice daily, substantial results should be found over a period of a few months.

To use the chart, attach it conveniently to a wall so that the central spot is in a line with the nose. Stand erect with the tip of the nose as close to the central spot as possible. Then move the eyes slowly clockwise, following the outer edge of each arm of the design, including the outermost spots, until returning to the point of beginning. Then repeat the same action in a counterclockwise direction.

Remember to breathe deeply and rhythmically. Breathing is essential to vision, as it oxygenizes the blood. Try to do this exercise using real sunlight, not artificial light. Natural sunlight is good for the eyes.

After exercising through a complete cycle, going in both directions, blink and relax the eyes. Do three to five minutes of Palm Healing on the eyes. Then repeat a complete cycle, being careful to avoid eyestrain.

Suggested Reading

Aihara, Herman. *Basic Macrobiotics*. Tokyo and New York: Japan Publications, Inc., 1985.

Bailentine, Rudolph, M.D. *Diet and Nutrition*. Himalayan International Institute.

Bates, W. H., M.D. *Better Eyesight without Glasses*. Pyramid Books, 1968.

Colbin, Annemarie. *The Book of Whole Meals*. Autumn Press, 1979. *East West Journal.*

Esko, Edward and Wendy Esko. *Macrobiotic Cooking for Everyone*. Tokyo and New York: Japan Publications, Inc., 1980.

Esko, Wendy. *Introducing Macrobiotic Cooking*. Tokyo and New York: Japan Publications, Inc., 1978.

Fulder, Stephen. *The Tao of Medicine*. New York: Destiny Books, 1982.

Hsu, H. Y. *Practical Introduction to Major Chinese Herbal Formulas*. Oriental Healing Arts Institute.

Huxley, Aldous. *The Art of Seeing*. California: Creative Arts Book Company, 1982.

Jensen, Bernard. *The Science and Practice of Iridology*. Bernard Jensen.

Kushi, Michio. *The Book of Macrobiotics*. Tokyo and New York: Japan Publications, Inc., 1977.

———. *How to See Your Health*. Tokyo and New York: Japan Publications, Inc., 1980.

Kushi, Michio and Aveline Kushi. *Macrobiotic Diet*. Edited by Alex Jack. Tokyo and New York: Japan Publications, Inc., 1985.

Lappé, Frances Moore. *Diet for a Small Planet*. New York: Ballantine Books, 1975.

Masunaga, Shizuto with Wataru Ohashi. *Zen Shiatsu*. Tokyo and New York: Japan Publications, Inc., 1977.

Mendelsohn, Robert, M.D. *Confessions of a Medical Heretic*. Chicago: Contemporary Books, 1979.

Muramoto, Naboru. *Healing Ourselves*. New York: Avon Books, 1973.

Ohsawa, George. *Zen Macrobiotics*. Los Angeles: The Ohsawa Foundation, 1965.

Ohsawa, George with Herman Aihara. *Macrobiotics: An Introduction to Health and Happiness*. The Ohsawa Foundation.

Ohsawa, Lima. *Macrobiotic Cuisine*. Edited by Phillip Janetta. Tokyo and New York: Japan Publications, Inc., 1984.

Oki, Masahiro. *Practical Yoga: A Pictorial Approach*. Tokyo and New York: Japan Publications, Inc., 1970.

———. *Zen Yoga Therapy*. Tokyo and New York: Japan Publications, Inc., 1979.

Sattilaro, Antony, M.D. with Tom Monte. *Living Well Naturally*. Boston: Houghton Mifflin, 1984.

Serizawa, Katsusuke, M.D. *Effective Tsubo Therapy*. Tokyo and New York: Japan Publications, Inc., 1984.

Teeguarden, Ron. *Chinese Tonic Herbs*. Tokyo and New York: Japan Publications, Inc., 1985.

Yamamoto, Shizuko. *Barefoot Shiatsu*. Tokyo and New York: Japan Publications, Inc., 1979.